# Graphene - A Wonder Material for Scientists and Engineers

*Edited by Mujtaba Ikram, Asghari Maqsood and Aneeqa Bashir*

Published in London, United Kingdom

Graphene - A Wonder Material for Scientists and Engineers
http://dx.doi.org/10.5772/intechopen.100684
Edited by Mujtaba Ikram, Asghari Maqsood and Aneeqa Bashir

Contributors
Haia Aldosari, Hosein Hasan Bouzari, Rasoul Malekfar, Laleh Farhang Matin, Nnditshedzeni Eric Nnditshedzeni Maluta, Lutendo Phuthu, Rapela Regina Maphanga, Gaurav Saini, Simranpreet Singh Gill, Sarabjeet Singh Sidhu, Yannick A.F. Kamta, Ravi Paul Gollapalli, Jeremy Reid, Tingyi Wei, Ali Roberto Ruiz-Hernández, Adrián Gutierrez-Cruz, Jessica Campos-Delgado, Halina V. Grushevskaya, George Krylov, Aster Woldu, Ramli Ramli, Rahmat Hidayat, Mujtaba Ikram, Sara Baig, Amna Batool, Aneeqa Bashir, Saadia Mumtaz, Muhammad Ikram, Mohsin Saeed, Khurram Shahzad, Muhammad Umer Farooq, Asghari Maqsood, Muazzam Ahmed, Hamza Asghar, Mahnoor Naeem, Shamim Aslam



First published in London, United Kingdom, 2023 by IntechOpen
IntechOpen is the global imprint of INTECHOPEN LIMITED, registered in England and Wales, registration number: 11086078, 5 Princes Gate Court, London, SW7 2QJ, United Kingdom

British Library Cataloguing-in-Publication Data
A catalogue record for this book is available from the British Library

Additional hard and PDF copies can be obtained from orders@intechopen.com

Graphene - A Wonder Material for Scientists and Engineers
Edited by Mujtaba Ikram, Asghari Maqsood and Aneeqa Bashir
p. cm.
Print ISBN 978-1-80356-431-9
Online ISBN 978-1-80356-432-6
eBook (PDF) ISBN 978-1-80356-433-3

For EU product safety concerns:
IN TECH d.o.o., Prolaz Marije Krucifikse Kozulić 3, 51000 Rijeka, Croatia,
info@intechopen.com or visit our website at intechopen.com.

# Meet the editors

Dr. Mujtaba Ikram has a BS (Hons) in computational physics, an MS in materials and surface engineering, and a Ph.D. in material sciences and engineering. His research interests include nanotechnology, renewable energy, materials science and engineering. His work has been cited by scientists from all over the world. He has authored/co-authored a number of publications with 170+ cumulative impact factor in such prestigious international journals as *Advanced Materials, Journal of Materials Chemistry C, Nanoscale Research Letters, RSC Advances, RSC New Journal of Chemistry, Chemcatchem, Journal of Alloys and Compounds, Applied Nanoscience, International Journal of Hydrogen Energy, Journal of Physics and Chemistry of Solids,* and *Journal of Solid State Chemistry*. His research has taken him to the USA, Italy, Egypt, Germany, Slovenia, China, Hong Kong, Malaysia, UAE and many other countries. He has attended research training/conferences/workshops on industrial physics, renewable energy, advanced carbon materials and nanotechnology in various parts of the world. He is a frequent visiting scholar at the Abdus Salam International Center for Theoretical Physics (ICTP), Italy. He was one of two young scientists from South Asia selected for TWAS Science Diplomacy, held in Trieste, Italy in 2013. He has frequently been invited as a lecturer by CAS-TWAS Beijing. He received a CAS-TWAS green technology award in 2015 and a CAS-TWAS green chemistry and technology award for his guest lectures in 2017. He has been awarded various prestigious fellowships, including CAS-TWAS presidential fellowship 2014, I-CAMP University of Colorado at Boulder (USA) fellowship 2012, ICTP Italy participant fellowship (thrice), UNESCO nano system workshop fellowship (Italy) 2013, inter-continental advanced materials and photonics participant fellowship, University of Cambridge (UK) 2013, Emerging Nations Science Foundation (ENSF) travel fellowship 2012 and NUST foreign research presentation grant 2012.

Professor Emerita Dr. Asghari Maqsood is an advisor to the Vice-Chancellor of Air University, Islamabad, Pakistan. She served as a founding Dean of the Faculty of Basic and Applied Sciences at Air University from July 2015 to November 2022. She has over 48 years of experience in advanced materials research. She obtained her MSc in materials science from Oxford University, UK, a Ph.D. in materials science from Goteborg University, Sweden and a diploma in experimental work from Uppsala University, Sweden. She has 285 research publications to her credit, including 214 journal publications and seven book chapters. She has received many national and international awards, including a Gold Medal and a Merit Certificate from the University of Punjab (1971-72), the INFAQ foundation prize and a Gold Medal from the Pakistan Academy of Sciences (2000), the President's Award and a cash prize for Pride of Performance, (2001), the Best University Teacher Award (2002) from HEC, Islamabad, the Prime Minister's Gold Medal and financial award on Salaam Teacher Day (2004), the President's Award, Izaz-i-Fazeelat for Academic Distinction (2005), a Civil Award Sitara-e-Imtiaz (2010), a PVS lifetime achievement award in recognition of her dedicated services in the field of vacuum science and technology (2013), Top Woman Physicist of Pakistan Award (2013), PCST publication award (2013), Dr. Z. A. Hashmi Gold Medal on World Science Day for Peace and Development (2014), Quaid-e-Azam University Golden Jubilee Award (2017). She was named among the top 2% of world scientists by Stanford University, USA in 2019 and 2021/22, as well as a lifetime achievement in the year 2019.

Dr. Aneeqa Bashir has been associated with the Physics Department at the University of Punjab since 2001 and is currently working as Assistant Professor. She is also an MS coordinator and student advisor in the department. She received her Ph.D. from Imperial College London, and both her MPhil in solid-state physics and MSc in physics from Punjab University, with distinction. She completed postdoctoral studies at Glasgow University, UK. She has over ten years of experience in advanced materials research and is a patent holder. She has authored/co-authored a number of publications with 150+ cumulative impact factor in such journals as Advanced Materials, Advanced Functional Materials, Applied Physics Letters, Journal of Applied Physics, ACS Omega, Energies, Coatings and Journal of Electronic Materials. She has been awarded research fellowships by the Schlumberger Foundation, the Higher Education Commission of Pakistan, the National Science Foundation, and ICTP, Italy. She has participated in and presented her work at international conferences in the UK, France, USA and Pakistan.

# Contents

# Preface

In the second part of the eighteenth century, carbon was identified as an element for the first time. "A.L. Lavoisier" introduced the name carbon in 1789 and derived from the Latin word "carbo" which signified "charcoal (or ember)" to the Romans. After hydrogen, oxygen, and helium, Carbon is 4th frequent element in our solar system (sun, stars, comets, and planets' atmospheres), the 6th most abundant prevalent element in the universe and in earth's crust, it is the 17th most extensive element. It is primarily found in hydrocarbons and other chemicals. Surprisingly after oxygen, carbon is found as 2nd abundant mineral in the human natural body, accounting for around 18% of a person's total weight. Although, carbon is in short supply on the earth's surface, with hardly 0.2% makes of the planet's whole mass. In spite of that, its function is extremely vital because it has the ability to establish links with other atomic nuclei and with itself. As a result, carbon's ability to catenation cleared the path for the advancement of chemistry (basis of organic chemistry), the rich diversity of structural forms of solid carbon, and biology, allowing for the emergence of life's marvels. Advanced carbon-based materials are considered the backbone of the coming generations' technological and scientific era especially in the divisions of materials-based industry, and nanotechnology, respectively. In recent times, advanced carbon-based materials are of significant highlight. Owing to efficient physical properties, from a technological perspective, it has been employed in the environment, energy, electronic and thermal applications, respectively. Due to the wide field of applications, twenty-first century is named as "Century of graphene". Graphene is considered an outstanding form of carbon owing to good electrical conductivity, suitable strength and efficient permeable properties. In the following chapters, we will review the basic forms, properties and significant structures of allotropes of carbon which will be beneficial from a technological perspective.

Because of its versatile chemical, physical, and electrical properties, graphene as well as its nanocomposites are regarded as the backbone of engineering and scientific innovation. Different physical and chemical methods are used to create sustainable carbon materials. Furthermore, fabrication methods are employed in order to produce the composites, which are of constituents with desirable properties. Because of their biocompatibility, graphene nanomaterials have enormous potential for improving biology and drug delivery. The proposed book provides a variety of fabrication methods for sustainable graphene composites and highlights various applications of graphene. Furthermore, graphene nanocomposites are promising multifunctional materials with improved tensile strength and elastic modulus. Despite some challenges and the fact that carbon nanotube/polymer composites are sometimes better in some specific performance, graphene nanocomposites may have a wide range of potential applications due to their outstanding properties and the low cost of graphene. Because these graphene composites have a controllable porous structure, a large surface area, high conductivity, high-temperature stability, excellent anti-corrosion properties, and composite compatibility, they can be used in energy storage as electrocatalysts, electro-conductive additives, intercalation hosts, and an ideal substrate for active materials. Meanwhile, the book summarizes the graphene nanocomposite requirements, computational and physical properties studies for technological innovation and scientific applied research.

**Mujtaba Ikram**
Institute of Chemical Engineering and Technology,
University of the Punjab,
Lahore, Pakistan

**Asghari Maqsood**
Department of Physics,
Air University,
Islamabad, Pakistan

**Aneeqa Bashir**
Department of Physics,
University of the Punjab,
Lahore, Pakistan

Chapter 1

# Introductory Chapter: Brief Scientific Description to Carbon Allotropes – Technological Perspective

*Sara Baig, Muazzam Ahmed, Amna Batool, Aneeqa Bashir, Saadia Mumtaz, Muhammad Ikram, Mohsin Saeed, Khurram Shahzad, Muhammad Umer Farooq, Asghari Maqsood and Mujtaba Ikram*

## 1. Introduction to carbon

In the second part of the eighteenth century, carbon was identified as an element for the first time. "A.L. Lavoisier" introduced the name carbon in 1789 and derived from the Latin word "carbo" which signified "charcoal (or ember)" to the Romans. After hydrogen, oxygen, and helium, Carbon is 4th frequent element in our solar system(sun, stars, comets, and planets' atmospheres), the 6th abundant prevalent element in universe and in earth's crust, it is 17th most extensive elemen [1]. It is primarily found as hydrocarbons and other chemicals [2, 3]. Surprisingly after oxygen, carbon is found as 2nd abundant mineral in the human natural body [4], accounting for around 18% of a person's total weight. Although, carbon is in short supply on earth's surface, with hardly 0.2% makes of planet's whole mass [1, 3, 5]. In spite of that, its function is extremely vital because it has the ability to establish links with other atomic nuclei and with itself. As a result, carbon's ability of catenation cleared the path for the advancement of chemistry (basis of organic chemistry) [6], the rich diversity of structural forms of solid carbon [7], and biology, allowing for the emergence of life's marvels [4, 8]. Advanced carbon based material are considered backbone in coming generation technological scientific era especially in the divisions of materials based industry, and nanotechnology, respectively. During recent times, advanced carbon based materials are of significant highlight. Owing of efficient physical properties, from technological perspective, it has been employed in the environmnet, energy, electronic and thermal applications, respectively. Due to wide field of applications, twenty-first century is named as "Century of graphene". Graphene is considered as outstanding form of carbon owing to good electrical conductivity, suitable strength and efficient permeabile properties. In following chapter, we will review basics forms, properties and significant Structures of allotropes of carbon which will be beneficial from technological perspective.


IntechOpen

## 2. Basic structure of carbon

Carbon is foremost exciting element in periodic table to chemists because of its electrical structure and atomic size [6].

### 2.1 Electronic and nuclei configuration of the carbon atom

Carbon's symbol is C, and its atomic number (Z) is 6, indicating that the neutral carbon atom (C-12 isotope) contains six protons and six neutrons in its nucleus and likewise, six electrons in its electronic shells. $1s^2\ 2s^2\ 2p^2$ is the electron configuration or the electronic arrangement in each orbital [2].

### 2.2 Atomic mass (atomic weight) of carbon

The basis for establishing the atomic mass unit is the element carbon. By definition, an atomic mass unit is 1/12th of the atomic mass of the (C-12) isotope. The "International Union of Pure and Applied Chemistry approved" this concept in 1961. The atomic mass unit is absolutely, incredibly small in comparison to the traditional definition of a mass: one gram requires $0.6022 \times 10^{24}$ amu (also known as Avogadro's number or $N_A$). Natural carbon has 98.89 percent 12C and only 1.11 percent of 13C. Consequently, the average carbon atom's atomic mass is 12.01115 amu.

### 2.3 Carbon's radius

Carbon's atomic radius is half the gap between two carbon atoms in a planar graphite lattice. When compared to other elements (H, Li, He, B, Be, O, N, F), carbon has one of the shortest radii. Carbon has seven different isotopes out of which three isotopes have been identified (12C, 13C, 14C). The isotope 12C, which possesses six neutrons, is by far the most abundant. There are four to ten neutrons in the others (10C to 16C). C-12 and C-13 are the most stable isotopes of carbon, meaning they do not dissolve or change structure spontaneously. The other five-carbon nuclei are radioactive, which means they decay naturally by emitting particles, which are maybe an electron ($\beta^-$) or a positron ($\beta^+$) and are formed by neutron splitting [2]. In the study of organic reactions, the radioactive isotope 14C is used as a tracer. The magnetic moment (spin V2) of 13C makes it an appropriate probe for NMR research. For the atomic mass unit, 12C is utilized as the reference definition because of its enormous abundance and an emerging power. It is defined as having the Relative Atomic Mass of 12 precisely. This definition now governs all other nuclei and molecular masses [9].

## 3. Carbon allotropes

Allotropy is the property of an element that allows it to exist in multiple physical states. Carbon can be divided mainly into two types of allotropes which are as shown in the **Figure 1**:

- Crystalline Carbon Allotropes
- Amorphous Carbon Allotropes

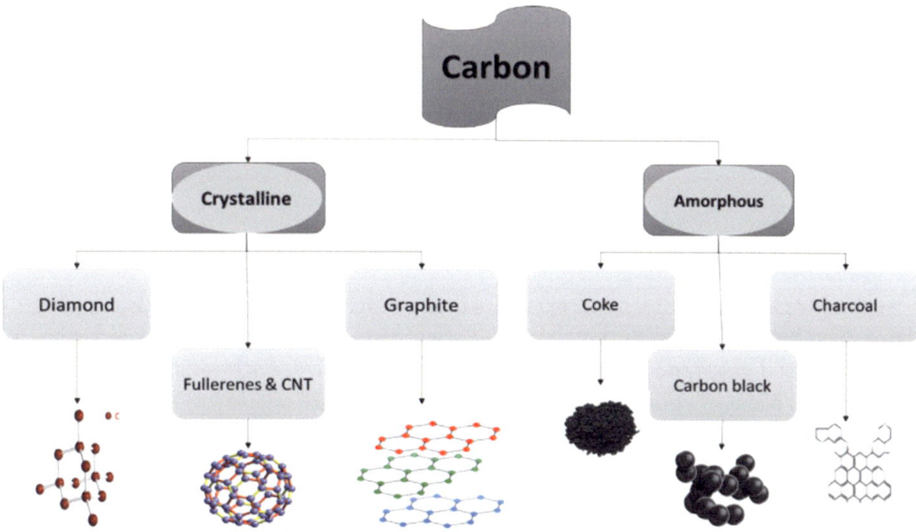


**Figure 1.**
*Flow chart of allotropes of carbon.*

Carbon is not the only element that has the ability to mix its atoms to produce allotropes. Silicon, germanium, and tin, all of which are found in the 4th column of the periodic table, share this property. Carbon, on the other hand, is distinctive in terms of its amount and allotrope's variety that it contains, due to its propensity to have variable oxidation states and coordination numbers [2].

### 3.1 Crystalline carbon allotropes

Carbon has been thought to have life, in mainly two crystalline allotropic forms for a long period of time: graphite and diamond, which are consist of extensive networks of $sp^3$ and $sp^2$ hybridized carbon atoms, respectively. As a result, we can say that diamond as well as graphite are the two most frequent allotropic forms of carbon. Both kinds have unique physical characteristics. In spite of that, the chemical composition of these two compounds differs significantly in terms of crystal shapes and characteristics [10–14]. Many other types of carbon allotropes may be produced chemically because of valence shell of the carbon atoms [15], which are chemically identical (same building block) but have dramatically different physical properties, such as buckminsterfullerene [16], also known as "Bucky-balls" [2, 17] carbon nanotubes [18], graphene [19, 20], and so on [8]. The discovery of fullerenes by "Harold W. Kroto "in 1985 signaled the start of an era of artificial carbon allotropes that included, in 1991 the production of (CNTs) and in 2004 again the discovery of graphene [20]. Therefore nowadays, carbon is significantly more than charcoal, because it produces the strongest fibers, most excellent lubricants (graphite), the strongest crystal and hardest substance (diamond), optimal gas absorbers (charcoal-gray), and perfect helium gas barriers [2].

### 3.2 Amorphous carbon allotropes

Amorphous carbons and carbons with mixed phases exist alongside to crystalline carbon allotropes [17]. Carbon without crystalline structure is referred to as amorphous form of carbon. Like all other glassy materials, some (specific short-range patterns) may be seen but not any (specific long-range pattern) of the atomic positions are observed. The majority of amorphous carbon is made up of tiny lattices of either graphite or diamond like carbon. This Non-crystalline carbon refers to coal, soot, or carbon black colloquially [17].

## 4. Overview of graphite and diamond

At the mesoscopic scale, a material's properties are dictated not only by its chemical bonding but also by its dimensions and structure. This is especially true in the case of carbon-based substances. Carbon is abundant in nature, with most of it in the form of composite. Most of them are natural chemicals that are needed to make artificial carbon and carbon-based products. Natural graphite and diamond are the only two allotropes of carbon found as minerals on Earth [2]. Despite the fact that both materials are relatively inert at normal temperatures and pressures, they can transition into one another when subjected to certain conditions [19, 21]. In the lowest energy state, carbon possesses four electrons in its valence shell, two in the 2 s sub-shell and two in the 2p sub-shell. The development of three types of sp. composite orbitals (sp, $sp^2$, and $sp^3$) depends on the amount of p orbitals (1 to 3) mingling with the s orbital. The bases of graphite and diamond are formed by carbon atoms with ($sp^2$ and $sp)^3$ composite orbitals, which may make 3 and 4 bonds with nearby carbon atoms, sequentially [17].

### 4.1 Graphite overview

Graphite was proposed by "Werner and D.L.G. Harsten", who derived it from the specific Greek 'γράφειν' (Graphein), which means "to write," and it was first used in pencils in 1594. "Abraham Gottlob Werner "was the first to name it in 1789 and is made up of large flat networks of carbon atoms layered on top of one another [17]. This carbon allotrope has a honeycomb layered structure and is made of flat two-dimensional layers of carbon atoms (sp2 hybridized). Each layer is specifically consist of the carbon atoms arranged in planar hexagonal rings with a carbon–carbon bond length of 141.5 pico-meters. A C−C covalent bond connects each C atom in each layer to three other C atoms [22].

## 5. Graphite is of two forms: α and ß.

- In α form, the layers are organized in ABAB... order, with the third layer directly above the first. Each layer, on average, divides two levels that are identically orientated. Within a layer, all carbon−carbon bonds are identical (a finding that is explained in terms of total pi-bonding).

- The ß form, on the other hand in nature, a less prevalent form of graphite is organized in ABCABCA... stacking, in which every 4th layer is identical. Microcrystalline graphite forms the basis for non-crystalline carbon variations.

- Three of the four carbon atoms create sigma bonds, while the fourth form a pi-bond. Graphite is extremely flexible but non-elastic because of the delocalization of the π –electrons [22]. It is also a conductor [19] in the basal plane and an insulator normal to the basal plane, with strong thermal and chemical inert characteristics [23].

- On the contrary, van der Waals forces slackly link the layers side by side. The layers are proficient to skate past each other because of inter-layer frailty generated by the attractive force of van der Waals interactions. This is the explanation, for graphite's soft and skiddy physical qualities, which make it excellent lubricant in generator and electric motors [8].

- The mineral (graphite) has a lustrous black color that is shiny and impenetrable (that is, it reflects light evenly without glitter or sparkle) [22].

- Carbon in the form of graphite is the most stable. Although, Graphite is difficult to melt. The strong molecular bonds that hold the carbon atoms together require a lot of energy to break [22]. As a result, in thermochemistry, it is utilized to define the heat of production of carbon composites as the standard state [2, 17].

### 5.1 Diamond overview

Foremost well-known allotropes of carbon is diamond. Diamond is a Greek word that means "transparent" and "invincible." The layer organization sequence in diamond is ABCABCA... There are four spatial orientations of carbon atoms due to the direction and perception of the tetrahedral axis, resulting in two tetrahedral and two octahedral (eight-faced) forms of diamond [17]. Each carbon atom in diamond is sp3 hybridized, forming covalent connections (bond lengths equal to 1.54 Å) with four additional carbon atoms in the tetrahedral structure's corners. Diamond's strength comes from its steady network of covalent connections and hexagonal rings [2].

- Diamond is unable to conduct any electricity owing to the limited mobility of electrons in the lattice pattern [24], but it exhibits maximum thermal conductivity [25] of roughly 2200 W/(mK), which is way five times higher as of copper [26, 27]. Diamond is commonly utilized, in the semiconductor technology to avert silicon and all other related semiconductors from overheating [13, 28] due to its excellent thermal conductivity [29]. The average electric breakdown of the diamond is fifteen times that of typical semiconductors, while the average holes mobility is five times that of common semiconductors [2].

However, it has a dielectric constant half that of silicon, and when tested, it is likely to exhibit 'negative electron affinity'. According to many studies, it is a 5.5 eV [30] broad bandgap semiconductor when doped to (n or p types).

- A diamond is a massive carbon atom molecule. It has unusual optical qualities. Because it is tied genetically to carbon family [31], it is considered also as biocompatible inside a live organism. Diamonds are transparent and colorless (transmitting light from the far-infrared to the ultraviolet). They are classified as glossy because they shimmer and reflect light. These characteristics make them desirable in jewelry. It is particularly neutron radiation resistant and has an extraordinarily high refractive index. The scientific uses include an optical sensor component, diamond anvil cells, and so on. it has a natural lubricity in the air that is comparable to Teflon [8].

- Diamond is supposed the hardest natural mineral ever known to human and It is insoluble in water. It has a somewhat high melting point. As a result, it's ideal for cutting tools. This makes it a great abrasive with outstanding polish and luster retention. No known naturally occurring chemical has the ability to cut (or even scratch) [17]. Any electronics device built of the diamond should be able to operate in harsh situations [32] due to its physicochemical features (rugged and 'hard'). All of the features listed above are natural outcomes of the crystal structure.

## 6. Detailed overview of graphene

"Graphene" [20] is just a single thin layer of carbon atoms separated from the overall graphite structure. In the ideal scenario, graphene is an perfect (2D) material because it is an atomically thick and thermodynamically stable monatomic sheet of carbon atoms organized in a honeycomb structure [33]. "Hanns-Peter Boehm" was the first to create the name "graphene" in 1962. The idea of graphite layers had been investigated before to this date [32]. In 1947, Wallace exploring the graphite layers which proves as a beginning point for better understanding the physical properties of the three-dimensional graphite. Several papers were published during the next few decades in an attempt to isolate or grow graphene. In 2004, "Andre Geim and Kostya Novoselov" have been succeeded in isolating a single sheet of 2-D graphene from the raw graphite [19, 20, 34]. Single layers can be separated from graphite and produced using traditional CVD or micromechanical cleavage techniques [17, 19]. Recently, Single graphene layers have been successfully generated utilizing a basic mechanical exfoliation of graphite using Scotch tape [20].

In Graphene, only the layer extensions and characteristics of edges might be different, making it the most fundamentally uniform material [20]. In other words, graphene is a fundamental basis [35] for all fullerene allotropic dimensionalities, and it tends to evolve into different sorts of structures with greater structural stability. It can be rolled into (0-D)fullerenes [36], wrapped into (1-D) carbon nanotubes (CNTS), and layered into (3-D) graphite (with 3.37A$^{0}$ separation distance) [37] and derivatives between layers [17, 34, 35]. Single-layer graphene (SLG) [38], Graphene Nano-platelets (GNPs) 100 nm to 100 μm [37], Graphene oxide (GO) (single-layer material with a high oxygen concentration), Reduced graphene oxide (RGO) (mono-layer form), and functionalized/chemically modified graphene are all examples of graphene. All of them are graphene-related compounds and can be classified as graphene family materials (GFMs) [39].

### 6.1 Brief lattice structure of 2-D graphene

Graphene is consist of two interpenetrating Bravais sub-lattices, which allows us to define a primitive unit cell, which is the simplest building block from which the graphene lattice may be constructed. Due to the hexagonal shape, the primordial cell encompassed by the two lattice vectors below includes two atoms as shown in **Figure 2**, one of type A and the other of type B, which represent the two triangular lattices. The size of the graphite primitive unit cell is determined by how individual graphene layers stack together to form the graphite crystal [40]. The lattice positions ( $\vec{R}$ ) are filled

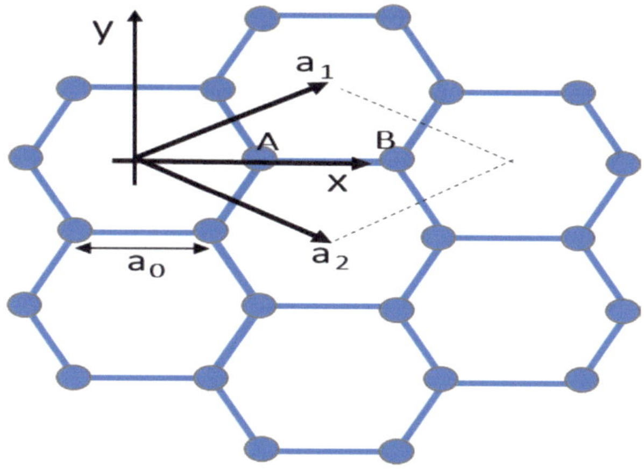


**Figure 2.**
*Single layer graphene lattice with two lattice vectors.*

with Type (A) atoms and the type (B) atoms are shifted with respect to the A atoms in each primitive cell by $\tau = (\vec{a}_1 + \vec{a}_2)/3$ [17].

$$\vec{a}_1 = \left(\frac{3}{2}a_o, -\frac{\sqrt{3}}{2}a_o\right),\ \vec{a}_2 = \left(\frac{3}{2}a_o, \frac{\sqrt{3}}{2}a_o\right),\ \vec{R} = m\vec{a}_1 + n\vec{a}_2.$$

Here, $a_o = 0.142nm$ is the bond length of carbon. Where $m$ and $n$ are integers.

### 6.2 Morphology of graphene

AFM, TEM, and FESEM can be used to study the morphology of graphene. AFM has measured the accurate thickness of single sheet of graphene to be (0.34–1.2 nm), and determining these thicknesses is crucial. Graphene has a higher thickness (2 nm) and the mean thickness of hydrazine-reduced GO, on the other hand, is only 0.8 nm, indicating the creation of single-layer graphene. The 'TEM and high-resolution TEM (HR-TEM)' are highly useful for detecting the number of layers in transparent graphene layers with crumples. The researchers looked into the microstructure variations between graphite and graphene and they found that Graphite layers are darker, thicker, and longer than graphene layers [41].

### 6.3 Physical properties of graphene

Graphene has attracted a lot of attention as a rising star in material science, solid-state physics, chemistry, and technology research because of its unique features, such as the quantum Hall effect, extraordinarily high (elasticity and tension), and optical transparency [17]. The 2p orbitals are responsible for graphene's amazing properties, as they cause the p bands to travel over the carbon sheets that make up the graphene. Graphene has a spectrum of remarkable features that other metals and semiconductors lack due to its unique lattice structure, shape, and surface morphology [17]. The material graphene is semimetal. Its zero bandgaps [37, 42], linear energy spectrum, excellent carrier mobility, frequency-independent absorption, and long spin diffusion length make it a popular material for electrical, photonic, and spintronic devices. Researchers discovered that graphene is highly rigid and extraordinarily good conductors of electricity and heat in the free-state, but are unstable at finite temperatures in the free-state [40], and is impermeable to gases especially [43]. Graphene is a semimetal with zero bandgap due to its Fermi level, which is located at the exact intersection of "conduction and valence bands" in pure substance and may be changed to make it (either N-Type or P-Type) by chemical modifications or more readily, by an electric field [44]. The inability to analyze the mechanical properties of single layer graphene is hampered by the fact that (SLG) is thermodynamically unstable and that an X-ray diffraction experiment on an (SLG) is not conceivable, owing to graphene's smallest size and outstanding surface-area-to-volume ratio. The tensile and compressive forces applied to the graphene lattice are tiny—around 1%. It's impossible to apply a substantially higher compressive strain without the sample slipping off the substrate (in the case of graphene). This can be accomplished by creating a monolayer graphene sample on a substrate having an aperture and pressing down on the graphene sample [40] with an (AFM) tip over the aperture. Graphene has been shown to endure a tensile strain of up to 20% using this method which will be beneficial from technological perspective.

## 7. Overview of carbon nanotubes (CNTS)

With its discovery in 1991, CNTS are carbon tubes (nanometer diameters range) also termed as "bucky-tubes", are significant contribution in technological nanosciences. CNTS are distinguished in "multi-walled CNTs (MW-CNT)" and "single-walled CNTS (SW-CNT), respectively. CNTS are synthesized using thermal decomposition, catalytic growth, and arc-ablation, respectively. Structure of graphite (crystalline) need to be well understood to visualize single walled carbon nanotubes. As graphite contains stacked-layers of hexagonally-arranged C-atoms with $sp^2$-configuration. Owing to inter-molecular forces, stcking layer give us single layer which is graphene. Single nanotube consist of C-atoms (millions) but thickness of tube is ~1 atom which may prove benefucial from technological perspective. The structures of SW-CNTS and MW-CNTS are shown in the **Figure 3**.

## 8. Conclusion

In the second part of the 18th century, carbon was identified as an element for the first time. "A.L. Lavoisier" introduced the name carbon in 1789 and derived from the Latin word "carbo" which signified "charcoal (or ember)" to the Romans. After hydrogen, oxygen, and helium, Carbon is one of abundant element in earth's crust,

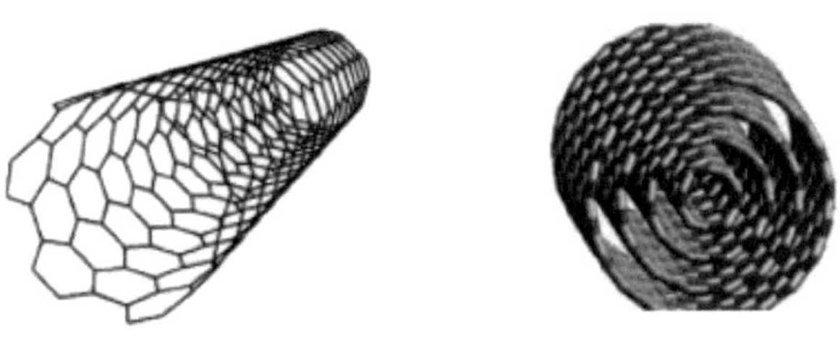

**Figure 3.**
*(a) SW-NT [8] and (b) MW-NTs [15].*

solar system, and universe. The chapter sums up basics properties of various allotropes of carbon along with its advanced forms. Advanced carbon materials, which includes graphene and carbon nanotubes, are prospective additions in materials based technology due to their outstanding structures and physical properties. Advanced carbon materials will be future of next scientific revolution in electronics, bio medical, energy and electrical engineering from technological perspective.

## Acknowledgements

The authors are gratified to University of the Punjab through research grant/Dr. Mujtaba Ikram/No. D/72/Est-l.

## Author details

Sara Baig[1], Muazzam Ahmed[2], Amna Batool[1], Aneeqa Bashir[2], Saadia Mumtaz[1], Muhammad Ikram[3], Mohsin Saeed[2], Khurram Shahzad[1], Muhammad Umer Farooq[4], Asghari Maqsood[5] and Mujtaba Ikram[1*]

1 Institute of Chemical Engineering and Technology (ICET), University of the Punjab, Lahore, Punjab, Pakistan

2 Department of Physics, University of the Punjab, Lahore, Punjab, Pakistan

3 Department of Physics, Government College University Lahore, Punjab, Pakistan

4 Department of Physics, Division of Science and Technology, University of Education, Lahore, Pakistan

5 Nanoscale Laboratory, Department of Physics, Air University, Islamabad, Pakistan

*Address all correspondence to: mujtaba.icet@pu.edu.pk

## References

[1] Zhang Y, Yin QZ. Carbon and other light element contents in the Earth's core based on first-principles molecular dynamics. Proceedings of the National Academy of Sciences of the United States of America. 2012;**109**(48):19579-19583

[2] Pierson HO, editor. 1 – Introduction and General Considerations, in Handbook of Carbon, Graphite, Diamonds and Fullerenes. Oxford: William Andrew Publishing; 1993 p. 1-10

[3] Allègre CJP, Humler E, Hofmann AW. The chemical composition of the earth. Earth and Planetary Science Letters. 1995;**134**:515-526

[4] Pace NR. The universal nature of biochemistry. Proceedings of the National Academy of Sciences of the United States of America. 2001;**98**:805-808

[5] Marty B, Alexander CMO, Raymond SN. Primordial origins of Earth's carbon. Reviews in Mineralogy and Geochemistry. 2013;**75**(1):149-181

[6] Braun T. Developments in Fullerene Science. Budapest, and Hungary: L. Eötvös University; 2002

[7] McEnaney B. CHAPTER 1 - Structure and Bonding in Carbon Materials, in Carbon Materials for Advanced Technologies. In: Burchell TD, editor. Oxford: Elsevier Science Ltd; 1999. p. 1-33

[8] Nasir S et al. Carbon-based Nanomaterials/allotropes: A glimpse of their synthesis: Properties and Some Applications. Materials (Basel). 2018;**11**(2)

[9] Edwards IAS, Chapter 1 - Structure in Carbons and Carbon Forms. In: Marsh H et al., editors. Introduction to Carbon Science. Butterworth-Heinemann. 1989. p. 1-36

[10] Ferrari AR. Interpretation of Raman spectra of disordered and amorphous carbon. Physical Review B. 2000;**61**:14095-14107

[11] Hodkiewicz JS, TF. Characterizing carbon materials with Raman spectroscopy. Progress in Materials Science. 2005;**50**: 929-961

[12] Titirici M. Sustainable Carbon Materials from Hydrothermal Processes. Chichester, UK: John Wiley and Sons, Ltd; 2013

[13] Wei L et al. Thermal conductivity of isotopically modified single crystal diamond. Physical Review Letters. 1993;**70**(24):3764-3767

[14] Dai L et al. Carbon nanomaterials for advanced energy conversion and storage. Small. 2012;**8**(8):1130-1166

[15] Pang J et al. CVD growth of 1D and 2D sp2 carbon nanomaterials. Journal of Materials Science. 2015;**51**(2):640-667

[16] Kroto HW, Health JR, O'Brien SC, Curl RF, Smalley RE. C60: Buckminsterfullerene. Nature. 1985;**318**:162-163

[17] Meeker K. Allotropes of carbon. 2016

[18] Iijima S. Carbon nanotubes: Past, present, and future. Physica B: Condensed Matter. 2002;**323**(1-4):1-5

[19] Potts JR et al. Graphene-based polymer nanocomposites. Polymer. 2011;**52**(1):5-25

[20] Hirsch, A., The era of carbon allotropes. Nat Mater. 2010.**9**(11): p. 868-71

[21] Pierson H.O., 3 - Graphite Structure and Properties, in Handbook of Carbon, Graphite, Diamonds and Fullerenes. In:

H.O. Pierson, Editor. Oxford: William Andrew Publishing; 1993; p. 43-69

[22] Dresselhaus MS, Kalish R. Ion implantation in diamond, graphite and related materials. Vol. 22. Springer Science & Business Media. 2013

[23] Deprez NM, McLachlan DS. The-analysis-of-the-electrical-conductivity-of-graphite-conductivity-of-graphite-powders-during. Journal of Physics D: Applied Physics. 1998;**21**:101-107

[24] Brandt NB, Chudinov SM, Ponomarev YG, editors. Semimetals Graphite and Its Compounds Compress. Amsterdam, The Netherlands: Elsevier; 1988

[25] Ekimov EA et al. Thermal conductivity of diamond composites sintered under high pressures. Diamond and Related Materials. 2008;**17**(4-5):838-843

[26] Wu Y et al. High-frequency, scaled graphene transistors on diamond-like carbon. Nature. 2011;**472**(7341):74-78

[27] Deneuville A. Electronic properties, devices and applications of diamond thin films. Academy of Sciences. 2000;**1**:81-90

[28] Hausmann BJM et al. Fabrication of diamond nanowires for quantum information processing applications. Diamond and Related Materials. 2010;**19**(5-6):621-629

[29] Shatskiy A et al. Boron-doped diamond heater and its application to large-volume, high-pressure, and high-temperature experiments. The Review of Scientific Instruments. 2009;**80**(2):023907

[30] Hausmann BJM et al. Diamond nonlinear photonics. Nature Photonics. 2014;**8**(5):369-374

[31] Walkert J. Optical absorption and luminescence in diamond. Reports on Progress in Physics. 1997;**42**:1606-1659

[32] Sun MJ, Cao X, Cao Z. Si(C≡C)4-based single-crystalline semiconductor: Diamond-like superlight and superflexible wide-bandgap material for the UV photoconductive device. ACS Applied Materials & Interfaces. 2016;**8**:16551-16554

[33] Slonczewski JC, Weiss PR. Band structure of graphite. Physical Review. 1958;**109**(2):272-279

[34] Novoselov KS. Nobel lecture: Graphene: Materials in the flatland. Reviews of Modern Physics. 2011;**83**(3):837-849

[35] Wallace PR. The band theory of graphite. Physics Review. 1947;**71**(622):7

[36] Novoselov KS, et al., Electric field effect in atomically thin carbon films. science. 2004;**306**(5696):666-669

[37] Phiri J, Gane P, Maloney TC. General overview of graphene: Production, properties and application in polymer composites. Materials Science and Engineering: B. 2017;**215**:9-28

[38] Brownson DACK, Banks DK, Craig E. An overview of graphene in energy production and storage applications. Journal of Power Sources. 2011;**196**(11):4873-4885

[39] De Marchi L et al. An overview of graphene materials: Properties, applications and toxicity on aquatic environments. Science of the Total Environment. 2018;**631-632**:1440-1456

[40] Proctor JE, Armada DM, Vijayaraghavan A. An Introduction to Graphene and Carbon Nanotubes. 2016

[41] Kadish KM, D'souza F. Handbook Of Carbon Nano Materials. Vol. 5-6. World Scientific. 2014

[42] Dreyer DRRRS. Bielawski CW. From conception to realization: An historial account of graphene and some perspectives for its future. Angewandte Chemie. 2010;**49**(49):9336-9344

[43] Verma D, Goh KL. Functionalized graphene-based nanocomposites for energy applications, in Functionalized Graphene Nanocomposites and their Derivatives. 2019. p. 219-243

[44] Wolf, EL. Applications of Graphene: An Overview. 2014

Chapter 2

# Graphene Based Functional Hybrids: Design and Technological Applications

*Hamza Asghar, Sara Baig, Mahnoor Naeem, Shamim Aslam, Aneeqa Bashir, Saadia Mumtaz, Muhammad Ikram, Mohsin Saeed, Khurram Shahzad, Muhammad Umer Farooq, Asghari Maqsood and Mujtaba Ikram*

## Abstract

Because of the versatile chemical, physical, and electrical properties, graphene as well as its nanocomposites are regarded as the backbone of engineering and scientific innovation. Different physical and chemical methods are used to create sustainable carbon materials. Furthermore, fabrication methods are employed in order to produce the composites, which are of constituents with desirable properties. Because of their biocompatibility, graphene nanomaterials have enormous potential for improving biology and drug delivery. The proposed chapter provides a variety of fabrication methods for sustainable graphene composites and highlights various applications of graphene. Furthermore, graphene nanocomposites are promising multifunctional materials with improved tensile strength and elastic modulus. Despite some challenges and the fact that carbon nanotube/polymer composites are sometimes better in some specific performance, graphene nanocomposites may have a wide range of potential applications due to their outstanding properties and the low cost of graphene. Because these graphene composites have a controllable porous structure, a large surface area, high conductivity, high temperature stability, excellent anti-corrosion properties, and composite compatibility, they can be used in energy storage as electrocatalysts, electro-conductive additives, intercalation hosts, and an ideal substrate for active materials. Meanwhile, the chapter summaries the graphene nanocomposites requirements for technological innovation and scientific applied research.

**Keywords:** graphene, physical properties, polymers, ceramics, technological applications

## 1. Introduction to graphene/polymer nanocomposites

Graphene, as well as its, distinguish variants, possess excellent characteristics. A remarkable achievement for them, because of these traits is their utility as fillers in the fabrication of composites. There are various factors that impart critically in the properties of graphene-polymer composites. They include dispersity of fillers, bonding between matrix and filler, the ratio of graphene filler to the polymer matrix, and quality. However, there are several fabrication approaches for undergoing modification in these factors in order to acquire desirable properties. In general, graphene comprises a very high aspect ratio, therefore it is capable of doing several modifications in properties of composites even at a very low percolation limit.

### 1.1 Fabrication of graphene/polymer composites

The most widely used methods for the incorporation of graphene into polymer matrices are, in situ polymerization, melt intercalation, and solution mixing. These three schemes are having schematic format as shown in **Figure 1**.

#### *1.1.1 In-situ polymerization*

Here the scheme, constitutes the mixing of a monomer and graphene fillers, particularly in the presence of a catalyst (s). The stage is then generally followed by polymerization initiation through heat or radiation [1]. The utility of epoxy in in-situ polymerization is an excellent example [2]. However, fabrication done by this scheme yields some polymer/graphene composites like graphene/polystyrene (PS) [1], graphene/polyaniline (PANI) composite [3], and graphene/silicone composites [4]. Significance of this method is the strong interaction between the filler and the polymer matrix. This leads to a rapid stress transfer, enabling a quick formation of the homogeneous dispersion. In addition, fillers constitute homogeneity in their distribution. High filler loading in polymer matrices is also achieved through In-situ polymerization. Meanwhile, the mixture becomes more viscous during polymerization, and the ongoing process suffers difficulties, causing ultimate hindrance for the loading fraction [5]. Usually solvent is the prime option in the current method most of time in some cases, but it extensively raises the need for the removal of solvent [6].

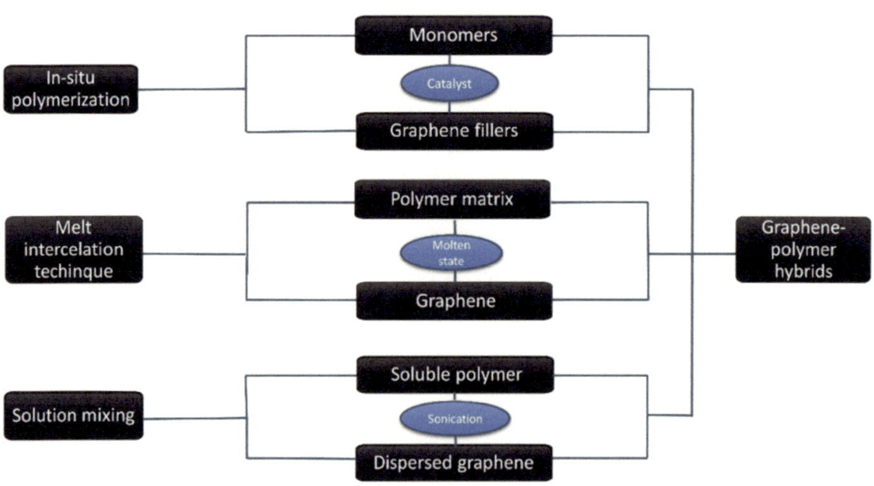


**Figure 1.**
*General fabrication routes for graphene polymer hybrids (GCHs).*

### *1.1.2 Melt intercalation technique*

Melt intercalation method incorporates designing of graphene in the molten state with the polymer matrix, thereby eliminating the need of solvent. High shear mixing with accelerated temperature results in the homogeneity of polymers and graphitic materials mixing. Thermoplastic composites are the most manufactured yield of this method. This method carries some of the graphene/polymer composites including exfoliated graphite/polypropylene nanocomposites [7], polystyrene/graphite nanosheet composites [8], polyethylene terephthalate (PET)/graphene nanocomposites [9], etc. On the contrary, this scheme has some discrepancies too, like poor dispersity and fillers distribution, than rest of methods. In addition, possibilities for breakage of graphene sheets and defects are comparatively much higher than rest of methods [10].

### *1.1.3 Solution mixing*

Fabrication of Polymer composites through solution mixing is the most frequent method. Potentially applicable on a bigger scale, simplicity and the demand of equipment with not much specifications are the main reasons for the common utility of this method. In this method graphene undergoes swelling as if s suited solvent refers solubilization of a polymer. This results in ultimate rise for the curvature of a surface. One of the most demanding parameter is graphene's dispersity and solubility as well. So, the need is functionalizing graphene sheets in a way ensuring ease for dispersion in various solvents. Some composites were having much higher solubility, that results in convenient fabrication like Poly(vinyl alcohol) (PVOH)/graphene oxide (GO) [11]. Sonication or shear mixing generally leads to a homogeneous mixture in many cases, as if the solvent is organic. There must be some definite schemes that alleviates aggregation while graphene sheets are functionalized [12].

## 1.2 Physical properties of graphene/polymer composites

Exfoliated carbon sheets are usually obtained through rapid pyrolysis or chemical reduction of graphene oxide. However, there are some ways that creates variation in the properties of these carbon sheets. In the following section, we would reveal some of the physical properties of graphene/polymer nanocomposites.

### *1.2.1 Thermal properties of graphene/polymer composites*

Vibrating lattice is the key upon which material's thermal conductivity depends on. Graphene ensures a material to be thermally conductive and stable, because it exhibits excellent thermal conductivity at room temperature ($\sim 3000 Wm^{-1}K^{-1}$). Therefore it imparts substantially in the improvement of thermal properties. Thermally less interface resistance, stiffness and the planar geometry of graphene are the major factors yielding remarkable thermal properties (1266 Epoxy/Graphite). Graphene's utility as transparent conductors attributes to its peculiar characteristics like higher conductivity, and incorporation towards ceramics and polymers [13]. On the contrary, thermal properties are suffered by graphene's orientation and aspect ratio.

*1.2.2 Mechanical properties of graphene/polymer composites*

Recent discoveries introduced fascinating insights of material science as graphene proved experimentally as nanomaterial. Graphene constitutes remarkable elastic modulus (125 GPa) and intrinsic strength ( $\sim 1.0TPa$ ). Scientists after having a comprehensive overview of all intrinsic properties of graphene, are now able to manufacture graphene-filled led polymer composites with desired properties. However, other milestones for the betterment include reinforcement phase aspect ratio, distribution in the host matrix, interface bonding, and the reinforcement phase concentration. Noticeable thing is that the increment in properties of polymer matrix attains at low filler loading [14].

*1.2.3 Electrical properties of graphene/polymer composites*

Electrical conductivity for graphene is its notable attribute, among all physical properties. Conductive graphene itself raises the possibility for elevation in electrical conductivity, when it comes to be utilized as fillers along with insulating polymer matrix. When loading fraction touches certain limit, particularly at a stage of percolation threshold, electrical conductivity increases rapidly right after the formation of network by filler. Electrical conductivity influences by the concentration of fillers as well as by percolation threshold. Conduction takes place through tunneling in polymer layers, so there is no such need for the filler to be in contact with the current flow directly. The limiting factor in thermal conductivity for polymer composites is ultimately the tunneling resistance [15]. Graphene in its other variant generally named as pristine graphene have much higher electrical conductivity, yet mechanical exfoliation scheme for its production on a bigger scale reduces its utility [16].

## 2. Introduction to graphene/semiconductor hybrids

Graphene semiconductor hybrids as well as composites prove to be more fascinated substitute for manufacturing comparatively much lighter, compact as well as effective batteries. If for instance, the concern is to alleviate various issues directing from synthesis routes to the recreatability of a homogeneous component, then the ultimate consequence would be the commercialization of products. Agglomeration rate of graphene matters too. Electrochemical sensors along with biosensors have upgraded their work by these hybrids. These sensors consider to be the best regarding detection of bacteria, viruses etc. [17].

### 2.1 Fabrication of graphene/semiconductor hybrids

ZnO combination with graphene reveals its speciality for better as well as increased performances. Fabrication of graphene semiconductor hybrids can be done by number of ways, like solvothermal method, ultrasonic spray pyrolysis, microwave-assisted zinc ions reduction with graphene as shown in **Figure 2**.

All these techniques possess limited utility because of some issues being unaddressed. These includes cumbersome procedure, equipment's sophistication and economically low feasibility. However, need is to adopt a method having no such issues. Therefore, In situ thermally decomposed zinc dihydrazinate complex on graphene's surface at comparatively moderate temperature is the best choice. Prime leverages for

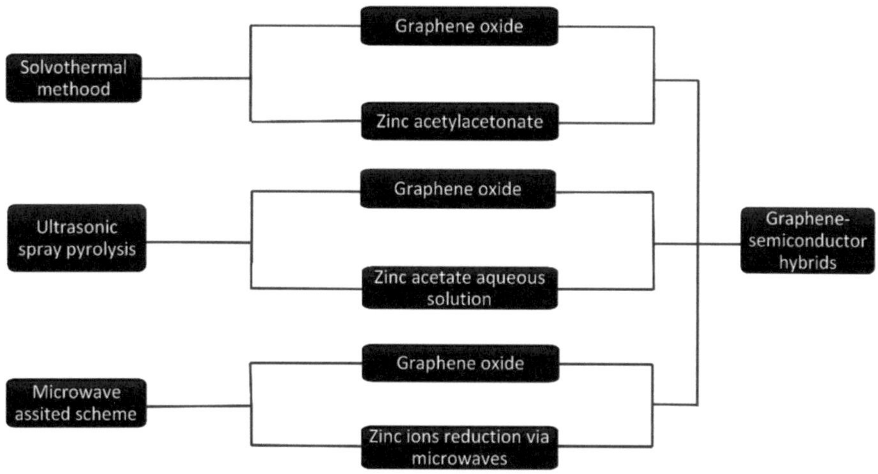


**Figure 2.**
*General fabrication routes for graphene semiconductor hybrids (GSH).*

this new method are its simplistic behavior, uniformity in size of particles, high end product with no much residual impurities [18].

*2.1.1 In-Situ thermal decomposition*

At first, aqueous solution of zinc (II) nitrate hexahydrate and hydrazinium benzoate undergoes through in situ mixing, thereby yielding a precursor complex. However, molar concentration of these two chemicals is 0.01 mol and 0.02 mol respectively. Both of them should be stirred consistently. However, the final mixture exhibits concentrated product comprising an alleviation in its volume. Then, all constituents should be subjected to drying after cooling, and cleaning process. Now take 1 mg of graphene and disperse it into 5 ml of ethanol, allowing it for stirring of 1 h, similarly 10 mg complex precursor of zinc benzoate dihydrazinate is to add up and repeat sonication process. The decomposition of complex into nanoparticles of ZnO results in their coating on graphene sheets as the transfer of contents to a fine silica pot, undergoes an environment of 200°C temperature for 2 h.

## 2.2 Physical properties of graphene/semiconductor hybrids

Graphene as well as Zinc oxide exhibits discrepancies up to some extent when considered separately. Therefore, the prime role of this combination mainly leads to the compensation of these flaws. This synergistic combination has been undergone comprehensive investigation, as it proves to be an excellent candidate of future devices, because of the enhancement in most of the properties like chemical stability, electrical etc. [19].

*2.2.1 Photocatalytic properties*

The interaction of graphene along with ZnO semiconductor, is ultimately vigorous. Therefore, it results in quickest movement of photo-generate negatively charged particles from Zinc oxide to graphene. Yet, another notable activity is the alleviation, for recombination process of these electrons. In general, activity of the photocatalytic process rises, thereby imparting a constructive role in these properties [19].

### *2.2.2 Electrical properties*

Another big advantage of graphene semiconductor hybrids is particularly its improvement of the electrode materials for supercapacitors. This renovation arises because of the various factors like specific surface area enhancement, outstanding cyclical stability, elevation in energy as well as power density, better ionic and electrical conduction performance. ZnO particles as well as graphene enclosed porous carbon have much improved conductivity of electrode entirely due to their combination. In addition, electron's movement from ZnO to the layer of graphene is the manifestation to the alleviation of defect emission. On the contrary, movement of charge from graphene to defect state of the given semiconductor causes an enhancement of defect levels contribution [19].

## 3. Introduction to graphene ceramic composites

Graphene is an excellent alternative ensuring the availability of composites having remarkable traits. Primarily its exceptional mechanical strength, results in potentially favorable supporting constituents for ceramic composites. Graphene is fascinatedly electrically conductive and this proves its versatility for various applications. Prime challenge for graphene's utility lies in its homogenous implantation into the ceramic matrices. Therefore, to address such challenges major milestone is its processing scheme. Herein the following sections will reveal number of processing schemes as shown in **Figure 3** for graphene based ceramic composites [20].

### 3.1 Fabrication of graphene/polymer composites

There are various segments upon the peculiar characteristic of these composites rely on. These includes fine particulates, equiaxed shape ensuring better packing, and homogeneous graphene's distribution into the ceramic matrix. However, another big challenge during investigation of toughening process, has been the fabrication of composites having well supervised micro/nano-structures [21]. Nowadays, appreciations for the sophisticated fabrication process are far more than conventionally adopted powder processing methods. These complex processing techniques comprise, colloidal processing scheme, sol–gel method, and polymer derived ceramic method as shown in **Figure 3**.

#### *3.1.1 Colloidal processing method*

Colloidal processing method exhibits intimate spread of ceramic as well as graphene matrix forming composites. These composites comprise microstructure that are homogenized and having properties ultimately controlled by colloidal chemistry. In colloidal processing technique, graphene is normally covered with ceramic fragments by the colloidal suspensions. This alters its surface chemistry, thereby alleviates repulsion between graphene, causing homogenous spread of graphene into ceramic grains. However, variation in surface chemistry consequences homogenous dispersion, that preserves itself even after sintering. Generally, for the demand of

*DOI: http://dx.doi.org/10.5772/intechopen.108791*

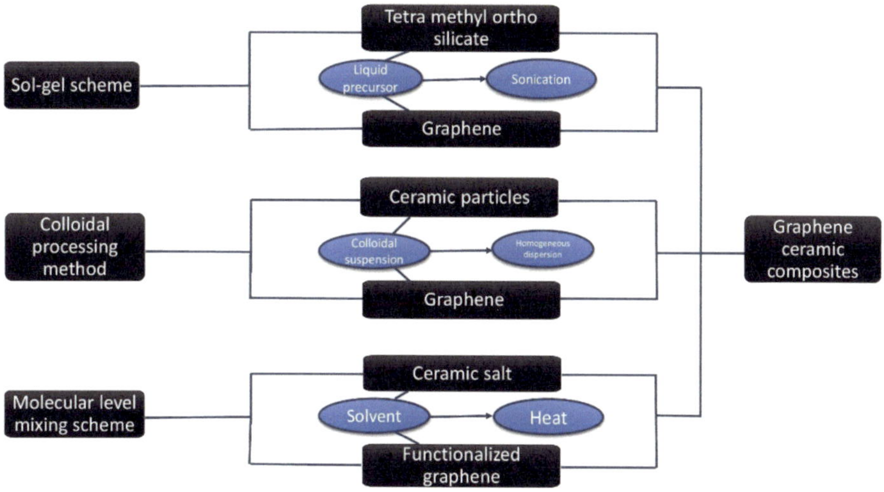


**Figure 3.**
*General fabrication routes for graphene ceramic composites (GCCs).*

homogenous spread, both these materials should be dissolved into the same kind of solvent. Although magnetic stirring is another key factor that causes slow mixing and imparts positively for the uniformity in spread of constituents [20].

### *3.1.2 Sol: gel method*

Sol–gel method comprises a substitute for the supply of dispersion of ceramic as well as graphene composites. Herein, graphene undergoes dispersion through a suitable molecular precursor ((TMOS) tetra methyl ortho-silicate), thereby leading to a condensation process for reinforcement. Homogeneous sol is readily available as the suspension of both the molecular precursor as well as graphene encounters sonication. In general, the method mainly ensures availability of nanocomposites of silica. Although, the technique suits well for fine spread, aggregation for precursor limits its utility [22]. However, sol–gel method need liquid precursors only, thereby providing ease for the processing of well-dispersed composites [23].

### *3.1.3 Molecular level mixing method*

Molecular level mixing scheme encapsulates another way for the production of ceramic and graphene composites. In this method, ceramic salt as well as functionalized graphene is combined in a suitable solvent. This then subjects its conversion to ceramic constituents via proper heat, thereby results in molecular level coating [24]. However, peculiar characteristic of this method includes excellent interface bonding between graphene and ceramic at molecular state. As a result of this strong bonding, a substantial enhance in the properties of the resulting composites becomes easier. Traditional methods of processing graphene-ceramic composites result in fall of mechanical properties. This occurs specifically owing to the effect of aggregation, caused by intermolecular forces [25]. Although, sol–gel method proves to be an excellent candidate for the homogenous dispersion of graphene into the ceramic matrix, however interfacial bonding between graphene and ceramic lacks stability [26].

## 3.2 Physical properties of graphene ceramic composite

Combination of graphene along with ceramic to form composites yields excellent physical properties, like mechanical, electrical and thermal.

### *3.2.1 Mechanical properties*

Ceramic graphene composites believe to be mechanically strong, when subjects to certain appropriate conditions. Yao et al. [27] analyzed that size of grains and fracture toughness are interconnected by each other. Their idea was that increasing the size more than 0.5 $\mu m$ predominately effects fracture toughness. Therefore, alleviation in fracture toughness of composites like AWTG2 as well as AWTG3 can be elaborated on the basis of the elevated growth in ceramic grains. Low concentration of ceramic composites results in number of pores, that results in substantial reduction of the interfacial bonding between graphene and ceramic grains [28]. However, graphene tends to pull-out because of this weak interfacial bonding, thereby enhances fracture toughness. This proves to be the ultimate cause for better fracture toughness of AWTG4 clears in (**Table 1**) [29].

### *3.2.2 Thermal properties*

Thermal conductivity suffers by various factors, and sintering at different pressure is one of them. At low temperature, thermal conductivity is proportional to the square of the temperature, reflecting amorphous like character. This happens because of various scattering phenomenon, especially phonons scattering by pores [30]. On the contrary, at high temperature conductivity rises at different rates, lightly for samples pressed by pressure of 5 GPa, while strongly for pressed at 4 GPa. For a high temperature range with increasing pressure, there is an unusual decrease in the value of thermal conductivity. Porosity as well as quality of graphene and ceramic composites are the of the main concern [31].

### *3.2.3 Electrical properties*

Electrical properties of graphene ceramic composites can be revealed on the basis of percolation theory as shown in **Figure 4**. Generally, in this theory threshold in percolation magnitude attributes to critical filler constituents, where increase in electrical conductivity results because of the presence of several conducting paths for electrons. These conducting paths of electrons do not stand if percolation magnitude lacks certain transition limit. Therefore, in order to ensure the availability of conducting chains in ceramics, percolation threshold must be much higher than threshold

| Sintered samples of (graphene-ceramic composites) | Fracture toughness |
|---|---|
| AWTG2 | 5.24 ± 0.17 |
| AWTG3 | 5.64 ± 0.21 |
| AWTG4 | 5.95 ± 0.23 |

**Table 1.**
*Sintered composites vs. fracture toughness.*

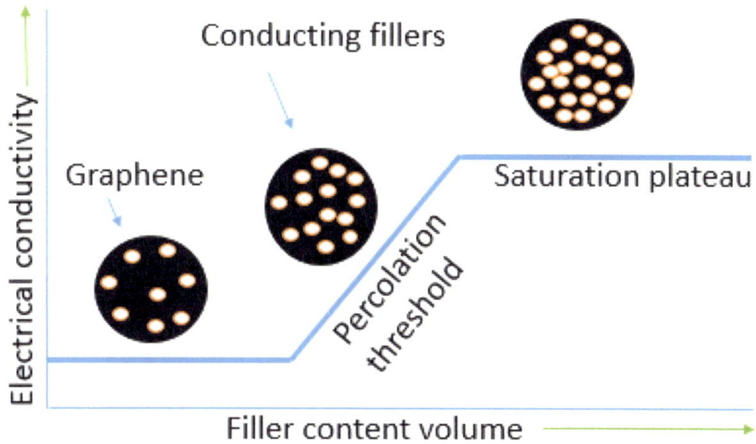


**Figure 4.**
*Electrical conductivity and percolation phenomenon as a function of filler volume fraction in graphene based ceramic composites.*

value. Hence, number of these electron paths causes an elevation in electrical conductivity of graphene ceramic composites [20].

## 4. Application of graphene

Comprehensive research in graphene's physical as well as chemical properties, has created a revolutionary in fields of electronics, energy storage devices, corrosion coating, etc. [32].

### 4.1 Drug delivery

Exploring various domains of graphene like biological one for enhancing its utility in biomedical field results as scientist's consistent effort in recent years. Free drugs leave defects intrinsically that comprises burst discharge, low blood circulatory span, irrelevant targeting. Therefore, it causes low therapeutic efficiency, lesser feasibility and variety of complexities [33]. Drug delivery server (DDS) is the ultimate need for alleviating these drawbacks. These DDS's have been used in various versions for last several decades as shown in **Figure 5**.

It simply encapsulates drug and commutes it to the desired destination, results as a controllable therapeutist [32].

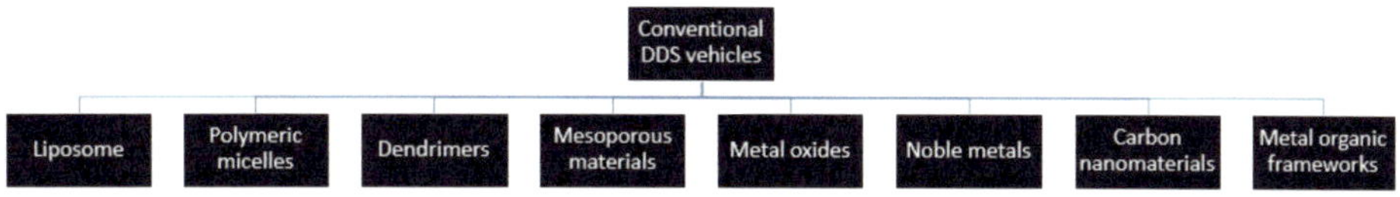


**Figure 5.**
*Traditionally used DDS vehicles from several decades.*

### 4.2 Electronics

Graphene's utmost demand for electronics results owing to its hike in the physical properties. High current capacity of transportation and various experimentally analyzed properties having numerical values are shown in **Figure 6** [34].

Radio-frequency transistors and analog electronics are ultimately the very finest tasks driven by graphene. It is because graphene comprises exceptional rise in its trans-conductance, material's stability and thinness. In addition, there is no need of switching off the devices entirely irrespective of its capability to do so, in radio-frequency transistors. For instance, in the process of signal's amplification in a single amplifier,the transistor is usually remains in the on condition [35]. Electro-absorption modulator is another utility as gate field intrinsically adjust graphene's Fermi level [36]. On-chip optical interconnections demand ultimately a very high bandwidth modulator, large speed and tiny footprints. Graphene having single layer causes slight light's absorption as the interaction of graphene with light is substantially rigorous. This problem is generally resolved by the coupling of graphene with a silicon waveguide. It creates an elevation by 0.1 dB $\mu m^{-1}$ of 1.35–1.60 $\mu m$ at frequencies greater than 1 GHz [36]. The notable benefit of graphene-based modulators is its sustainability in integrated form along with Si-CMOS electronics. One of the most dominant edge of graphene is its saturated absorption, that reveals particularly decrease of absorption light as a function of increase in light intensity. Saturable absorbers therefore, helps to turn continuous wave output into ultrashort light pulses. Picosecond laser pulses can be generated by graphene due to its peculiar traits like higher stability, quick decay and a broad absorption range [37].

### 4.3 Flexible and stretchable display

Flexible and stretchable display is the utmost need of the future electronics, and graphene is doing its job elegantly in this domain [38]. Beside remarkable properties, mineral resources consider so far unsuccessful for rigid components, and this limits their utility. On the contrary graphene and its variants are emerging candidates in flexible as well as in stretchable electronics [39]. Various distinguished traits ensures its availability in modern devices like LEDs and for conversion and in energy storage devices as well as shown in **Figure 7**.

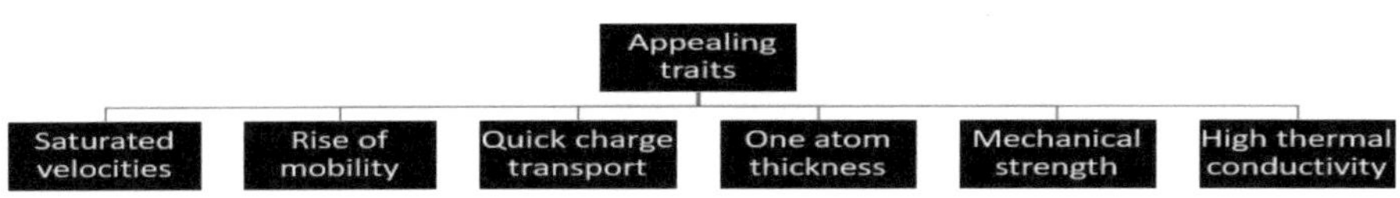


**Figure 6.**
*Fascinating characteristic of graphene in electronics.*

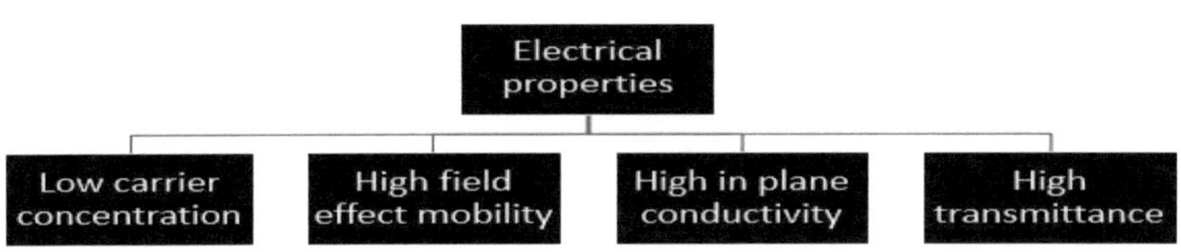


**Figure 7.**
*Excellent characteristic for flexible and stretchable displays.*

On the contrary, most utilized material like Indium tin oxide, exhibits wide range of discrepancies such as high corrosion rate, much higher refractive index, and less economic feasibility etc. [40].

### 4.4 Energy storage devices

Application for energy storage devices also attributes to graphene. This is due to the Ultimate reduction in restacking of graphene sheets specifically when interacting with metal oxides. The prime logic for graphene's utility in this arena of energy storage devices is its highly conductive nature, especially when it comes to interlinked network. Mechanically as well as electrochemically stable along with porosity in the structure all proves its significance in this current domain of application. Photovoltaic cells, batteries and fuel cells entirely rely on energy storage devices. For supercapacitors, ensuring material for electrode is graphene's job. The process of dysenetization can also be performed by these electrode material [41].

### 4.5 Corrosion coating

Graphene inherently behaves as a corrosion resistant, whenever coated. Penetration power constitutes a substantial increase in polymer coating. However, graphene enters along thickness of coating into the medium, and anti-corrosion process initiates right after the entrance of graphene to the interface. Carbon atoms in graphene exhibits sp2-hybridization, comprising high electron density, that renders all incoming corrosion molecules [22]. Hence, such kind of fascinating structure in graphene leads to its durability. In addition, graphene holds an excellent protection shield in the infiltration stage due to its capability of prolonging the path. Therefore, durability as well as corrosion resistance can be enhanced remarkably.

## 5. Applications of graphene based hybrids

Graphene based hybrids have a major role in food microbiology, where the microbes are applied on the Nano-scale level to prevent the packaged food from being rotten and it gives the food a long lifetime. Graphene based hybrids have been used in nano robotics. Nano-robotics is an emerging technology that creates machines or robots whose components are at or near the scale of a nanometer. Moreover, graphene based hybrids have been extensively used in drug delivery, food sciences, nano-optics, nano-energy, photo-voltaic, nano-chips, light-weight mechanical and electronics applications. Recently, Billions of dollars are being spent to research the use of nano-materials for the detection and treatment of diseases like chemotherapy and the manufacturing of minuscule sensors. Miniscule sensors can be injected into our body which can monitor our vitals even more closely than a doctor. Graphene based sensors are one of prominent addition in recent times for detections of diseases. Graphene based hybrid sensors can be manufactured in large quantities at a very low cost. These sensors are used to monitor the health of frameworks such as bridges and nuclear reactors as well as in the industries.

In addition, graphene-polymers based materials are trying to replace conventional materials in food packaging. And Nano-sensors are also developed to detect the presence of contaminants and dirt in the food particles respectively.

Nanotechnology provides a range of options to improve food quality and helps in improving food quality taste. Nano-encapsulation techniques have been used to improve the flavor of food.

## 6. Conclusion

Because of the unique mechanical, thermal, chemical, optical, and electrical properties, graphene and its nanocomposites have gotten a lot of attention in scientific research in recent years. Graphene with its nanomaterial, such as graphene polymer composites, graphene semiconductor composites and graphene ceramic composites are regarded as the foundation of material science and technological innovation. These nanomaterials are created using various physical and chemical methods to produce high-quality materials with excellent properties. Graphene nanomaterials are also used in electronics, organic photovoltaics, energy conservation technology, and drug delivery, among other things. In the future, these advanced materials could be used to create a variety of materials with various applications. A great deal of research is being conducted in order to produce these materials on an industrial scale. Because of their ability to store energy on a large scale, these materials are the future of sustainable energy production and storage devices. Fuel cells are also expected to replace battery-powered energy systems in the near future. Graphene sheets have the potential to revolutionize microelectronics. The demand for graphene nanocomposites will continue to rise as a result of technological advancement.

## Acknowledgements

The authors are gratified to University of the Punjab through Punjab University Research grant/Mujtaba Ikram for fiscal year 2022-2023.

*DOI: http://dx.doi.org/10.5772/intechopen.108791*

## Author details

Hamza Asghar[1], Sara Baig[2], Mahnoor Naeem[2], Shamim Aslam[1], Aneeqa Bashir[1], Saadia Mumtaz[2], Muhammad Ikram[3], Mohsin Saeed[1], Khurram Shahzad[2], Muhammad Umer Farooq[4], Asghari Maqsood[5] and Mujtaba Ikram[2*]

1 Department of Physics, University of the Punjab, Lahore, Pakistan

2 Institute of Chemical Engineering and Technology (ICET), University of the Punjab, Lahore, Pakistan

3 Department of Physics, Government College University, Lahore, Pakistan

4 Department of Physics, Division of Science and Technology, University of Education, Lahore, Pakistan

5 Nanoscale Laboratory, Department of Physics, Air University, Islamabad, Pakistan

*Address all correspondence to: mujtaba.icet@pu.edu.pk

## References

[1] Zheng W, Lu X, Wong SC. Electrical and mechanical properties of expanded graphite-reinforced high-density polyethylene. Journal of Applied Polymer Science. 2004;**91**(5):2781-2788

[2] Ramanathan T et al. Graphitic nanofillers in PMMA nanocomposites—An investigation of particle size and dispersion and their influence on nanocomposite properties. Journal of Polymer Science Part B: Polymer Physics. 2007;**45**(15):2097-2112

[3] Wang D-W et al. Fabrication of graphene/polyaniline composite paper via in situ anodic electropolymerization for high-performance flexible electrode. ACS Nano. 2009;**3**(7):1745-1752

[4] Verdejo R, Barroso-Bujans F, Rodriguez-Perez MA, De Saja JA, Lopez-Manchado MA. Functionalized graphene sheet filled silicone foam nanocomposites. Journal of Materials Chemistry. 2008;**18**(19):2221-2226

[5] Verdejo R, Bernal MM, Romasanta LJ, Lopez-Manchado MA. Graphene filled polymer nanocomposites. Journal of Materials Chemistry. 2011;**21**(10):3301-3310

[6] Du J, Cheng HM. The fabrication, properties, and uses of graphene/polymer composites. Macromolecular Chemistry and Physics. 2012;**213**(10-11):1060-1077

[7] Kalaitzidou K, Fukushima H, Drzal LT. A new compounding method for exfoliated graphite–polypropylene nanocomposites with enhanced flexural properties and lower percolation threshold. Composites Science and Technology. 2007;**67**(10):2045-2051

[8] Chen G, Wu C, Weng W, Wu D, Yan W. Preparation of polystyrene/graphite nanosheet composite. Polymer. 2003;**44**(6):1781-1784

[9] Zhang H-B et al. Electrically conductive polyethylene terephthalate/graphene nanocomposites prepared by melt compounding. Polymer. 2010;**51**(5):1191-1196

[10] Dasari A, Yu Z-Z, Mai Y-W. Electrically conductive and super-tough polyamide-based nanocomposites. Polymer. 2009;**50**(16):4112-4121

[11] Xu Y, Hong W, Bai H, Li C, Shi G. Strong and ductile poly (vinyl alcohol)/graphene oxide composite films with a layered structure. Carbon. 2009;**47**(15):3538-3543

[12] Phiri J, Gane P, Maloney TC. General overview of graphene: Production, properties and application in polymer composites. Materials Science and Engineering: B. 2017;**215**:9-28

[13] Potts JR, Dreyer DR, Bielawski CW, Ruoff RS. Graphene-based polymer nanocomposites. Polymer. 2011;**52**(1):5-25

[14] Huang X, Qi X, Boey F, Zhang H. Graphene-based composites. Chemical Society Reviews. 2012;**41**(2):666-686

[15] Lv C, Xue Q, Xia D, Ma M. Effect of chemisorption structure on the interfacial bonding characteristics of graphene–polymer composites. Applied Surface Science. 2012;**258**(6):2077-2082

[16] Verma D, Gope P, Shandilya A, Gupta A. Mechanical-thermal-electrical and morphological properties of graphene reinforced polymer composites:

A review. Transactions of the Indian Institute of Metals. 2014;**67**(6):803-816

[17] Ramírez C, Belmonte M, Miranzo P, Osendi MI. Applications of ceramic/graphene composites and hybrids. Materials. 2021;**14**(8):2071

[18] Kavitha T, Gopalan AI, Lee K-P, Park S-Y. Glucose sensing, photocatalytic and antibacterial properties of graphene–ZnO nanoparticle hybrids. Carbon. 2012;**50**(8):2994-3000

[19] Jana A, Scheer E. Study of optical and magnetic properties of graphene-wrapped ZnO nanoparticle hybrids. Langmuir. 2018;**34**(4):1497-1505

[20] Markandan K, Chin JK, Tan MT. Recent progress in graphene based ceramic composites: A review. Journal of Materials Research. 2017;**32**(1):84-106

[21] Wu Y, Kim G-Y. Carbon nanotube reinforced aluminum composite fabricated by semi-solid powder processing. Journal of Materials Processing Technology. 2011;**211**(8):1341-1347

[22] Cho J, Boccaccini AR, Shaffer MS. Ceramic matrix composites containing carbon nanotubes. Journal of Materials Science. 2009;**44**(8):1934-1951

[23] Zheng C, Feng M, Zhen X, Huang J, Zhan H. Materials investigation of multi-walled carbon nanotubes doped silica gel glass composites. Journal of Non-Crystalline Solids. 2008;**354**(12-13):1327-1330

[24] Hwang J et al. Enhanced mechanical properties of graphene/copper nanocomposites using a molecular-level mixing process. Advanced Materials. 2013;**25**(46):6724-6729

[25] Wang K, Wang Y, Fan Z, Yan J, Wei T. Preparation of graphene nanosheet/alumina composites by spark plasma sintering. Materials Research Bulletin. 2011;**46**(2):315-318

[26] DiMaio J et al. Transparent silica glasses containing single walled carbon nanotubes. Information Sciences. 2003;**149**(1-3):69-73

[27] Yao W et al. Grain size dependence of fracture toughness for fine grained alumina. Scripta Materialia. 2011;**65**(2):143-146

[28] Bódis E et al. Toughening of silicon nitride ceramics by addition of multilayer graphene. Ceramics International. 2019;**45**(4):4810-4816

[29] Wang X, Zhao J, Cui E, Liu H, Dong Y, Sun Z. Effects of sintering parameters on microstructure, graphene structure stability and mechanical properties of graphene reinforced Al2O3-based composite ceramic tool material. Ceramics International. 2019;**45**(17):23384-23392

[30] Stephens R. Low-temperature specific heat and thermal conductivity of noncrystalline dielectric solids. Physical Review B. 1973;**8**(6):2896

[31] Głuchowski P et al. Preparation and physical characteristics of graphene ceramics. Scientific Reports. 2020;**10**(1):1-10

[32] Song S et al. Biomedical application of graphene: From drug delivery, tumor therapy, to theranostics. Colloids and Surfaces B: Biointerfaces. 2020;**185**:110596

[33] Wang L, Zheng M, Xie Z. Nanoscale metal–organic frameworks for drug delivery: A conventional platform with new promise. Journal of Materials Chemistry B. 2018;**6**(5):707-717

[34] Liu G, Rumyantsev S, Shur M, Balandin AA. Graphene thickness-graded transistors with reduced electronic noise. Applied Physics Letters. 2012;**100**(3):033103

[35] Avouris P, Xia F. Graphene applications in electronics and photonics. MRS Bulletin. 2012;**37**(12):1225-1234

[36] Liu M et al. A graphene-based broadband optical modulator. Nature. 2011;**474**(7349):64-67

[37] Zhang H, Bao Q, Tang D, Zhao L, Loh K. Large energy soliton erbium-doped fiber laser with a graphene-polymer composite mode locker. Applied Physics Letters. 2009;**95**(14):141103

[38] Han T-H, Kim H, Kwon S-J, Lee T-W. Graphene-based flexible electronic devices. Materials Science and Engineering: R: Reports. 2017;**118**:1-43

[39] Kim SJ, Choi K, Lee B, Kim Y, Hong BH. Materials for flexible, stretchable electronics: Graphene and 2D materials. Annual Review of Materials Research. 2015;**45**(1):63-84

[40] Bonaccorso F, Sun Z, Hasan T, Ferrari A. Graphene photonics and optoelectronics. Nature Photonics. 2010;**4**(9):611-622

[41] Olabi AG, Abdelkareem MA, Wilberforce T, Sayed ET. Application of graphene in energy storage device – A review. Renewable and Sustainable Energy Reviews. 2021;**135**:110026

Chapter 3

# Graphene Reinforced Polymer Matrix Nanocomposites: Fabrication Method, Properties and Applications

*Haia Aldosari*


## Abstract

Graphene has exceptional mechanical capabilities, making it a potential reinforcement material for polymer composites. It also has unique electrical and thermal properties, making it an appealing filler for multifunctional composites, particularly polymer matrix composites, due to its vitality and superior mechanical qualities. This chapter thoroughly examines current graphene research trends, focusing on graphene-based polymer nanocomposites, manufacturing, characteristics and applications. Graphene-based materials are single- or multi-layer platelets that may be mass produced using chemical, physical and mechanical processes. A range of technologies for producing graphene-based materials, as well as methods for dispersing these nanoparticles in different polymer matrices, are being examined. The electrical, mechanical and thermal properties of these nanocomposites are also discussed, as well as how each of these features is influenced by the inherent properties of graphene-based materials and their state of dispersion in the matrix. It follows with a review of graphene's effect on composites and the difficulty of satisfying future industrial requirements.


**Keywords:** composites, graphene oxide, reduced graphene oxide, pristine graphene, polyethylene, polypropylene, thermal stability

## 1. Introduction

Graphene-reinforced polymer is classified as a multiphase material containing a single type of polymer, copolymer or a blend of polymers with nanofillers or nanoparticles (with dimensions of 1–50 nm) incorporated into the polymer matrix. This considerably affects the different physical, chemical and mechanical properties.

The plurality of the study has concentrated on polymer nanocomposites based on nanofillers: pristine graphene (G), reduce graphene oxide (rGO) and graphene oxide (GO), intending to improve the polymer's electrical, mechanical, thermal and gas barrier properties [1, 2]. Recently, graphene has shown the greatest promise as a nanofiller due to its superior exceptional physical properties. This has created a novel category of polymeric nanocomposites. Graphene, a novel type of carbon, is a one-atom thick plane in a two-dimensional sheet formed of $sp^2$ hybridized carbon atoms arranged in a hexagonal crystalline structure. It is the thinnest recognized material at present [3, 4]. In addition, it is one of the distinct allotropes of carbon that is the basic block to building all

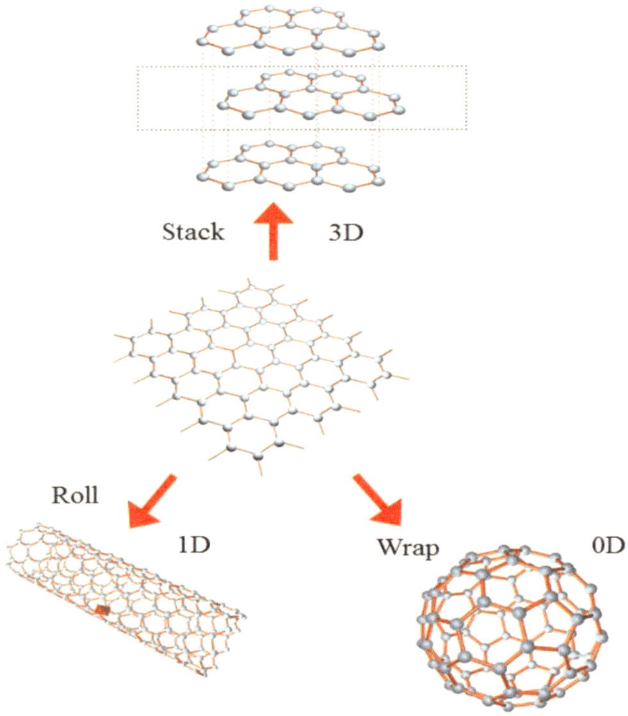


**Figure 1.**
*Source of all graphitic forms. Graphene is a 2D building material for carbon materials of all other dimensionalities. It can be made into Buckyballs, 1D nanotubes or to 3D graphite [5].*

graphitic derivative forms shown in **Figure 1**. Graphene can be arranged and stacked in each layer into graphite with three dimensions (3D), rolled into carbon nanotubes with one dimension (1D) and wrapped into fullerene with zero dimensions (0D). Graphene with two dimensions (2D) has distinct physical, chemical and engineering properties, with a large surface area, high thermal stability, electrical and thermal conductivity and high stiffness. These unique features make graphene a promising nanofiller in the field of polymer nanocomposites. As well it exhibits great potential for many applications in different fields such as electronic, medical and engineering [6–10]. Polyolefin (PO) nanocomposites based on nanofillers offer many opportunities to improve and develop the POs, with just small loud amounts of nanofillers. Recently, graphene has been explored for use as a promising nanofiller for POs. Many published articles demonstrate that graphene can be used for the reinforcement of polyolefins due to its exceptional physical and mechanical characteristics [11, 12]. The polyolefin/graphene nanocomposite is still in the early steps of development and improvement. However, the enormous possibilities of this material have become obvious in different research fields including automotive, electronics, and recently, gas and water barrier applications. The main challenge to completely exploiting graphene/polyolefin composites is to achieve a high level of homogeneous dispersion of graphene for the maximum benefit [13].

## 2. Introduction to graphene-reinforced polymer nanocomposites

Nanotechnology is used in many fields with applications ranging widely from medical to construction. The unique feature of this technology is its size. Materials with nano size have distinct characteristics such as a high surface area with low surface defects, which impacts significantly upon the characteristics of the consequent material. To illustrate, in nanotechnology, composites can be used as materials filler to decrease the weight of composite and increase the composite stiffness and fire resistance.

Nanocomposites are extensively used in different applications, e.g. solar cells, transport, construction and several other new implementations because of their unusual properties. They present superior mechanical and thermal properties, whilst being lightweight, characteristics which are complicated to obtain separately from the parent components. Nanocomposites, compared to classic composites, have a nanosize dimension and an exclusive set of characteristics because of their nano size. Consequently, this modern type of material presents progressive technological opportunities. Recently, a significant research body has focused on polymer nanocomposites both in the engineering and scientific fields to explore the distinctive properties of the nanosize system. It offers a sustainable alternative to classical loaded polymers, by adding nanofillers that have a high surface area to a polymer host matrices substance. The poor performance of most polymers can be enhanced to meet the needs and requirements of a wide range of scientific and engineering applications. In polymer nanocomposites, various categories of polymers, such as thermoplastics, thermosets and elastomers can be used as materials matrices. However, thermoplastic-based nanocomposites are attracting the most attention from both academic and industrial sources, due to their potential to be recyclable. The thermomechanical recycling procedure is the most cost-effective process for large scales of polymers. Throughout the thermomechanical recycling process, polymers undergo several kinds of thermal and mechanical processes that could change the polymer molecular structure, consequently changing the polymer performance. Recycled polymers usually have lower performance compared to original polymers, especially in applications that require low-strength polymers. The added nanofillers such as graphene have the potential to improve the properties of the polymer even after recycling [14].

## 2.1 Nanotechnology

Nanotechnology refers to materials and devices with design, characterization, production and application at a nanometer scale. Nano is a Greek word that means dwarf, indicating a decrease of size or time, $10^{-9}$ fold, that is smaller than a micron by 1000 times. One cubic nanometer ($nm^3$) is approximately 20 times the volume of an individual atom. A nanoelement's size relative to a basketball is the same as a basketball's size compared to the earth. These nanoscale materials display at least one unique feature because of their nanoscale size. A high-surface area and quantum effects from the nanosize material contribute to improving the materials by reinforcing their reactivity, and thermal, electrical and mechanical properties. Nanoscience studies the structure and properties of materials at atomic and molecular levels, based on the dimensions of the materials [15].

## 2.2 Nanoparticles

Particles with one dimension at least, that is around 1000 nm (1 micron) and less, and possibly as atomic size and molecular length scales (~0.2 nm), are termed nanoparticles. Nanoparticles can take both crystalline and amorphous forms and have a high-surface area per unit of volume. That unique property offers greater chemical reactivity than any other particles with a larger size, even with the same surface. To a certain degree, nanoparticulate material should be considered a featured state of the material, in addition to solid, liquid, gas and plasma states, because of its unique features with a large surface area. Typical nanoparticle crystalline forms are fullerenes and carbon nanotubes, while conventional crystalline forms are graphite and diamond. The materials formed from nanoparticles offer unfamiliar characteristics compared to conventional bulk materials. Many researchers limit the size of nanomaterials to around 50 nm [16] or 100 nm [17]. This maximum limit is justified by the actuality that some physical properties of nanoparticles equal those of bulk particles when their size reaches these values. However, a fair definition extends this upper limit, so that many particles up to 1 micron are classified as nanoparticles [18].

## 2.3 Nanomaterials

Materials that contain structural building blocks of less than 1 μm and at least zero dimensions are known as nanomaterials. However, bulk crystals with nanometer lattice spacing but macroscopic dimensions overall are generally eliminated.

### *2.3.1 Nanomaterials classification*

Nanomaterials are categorized based on many features such as nanoparticle geometry, morphology, composition, uniformity and agglomeration. This research uses a rating based on nanoparticle dimensions, whereas nanomaterials are classified as 0D, 1D, 2D and 3D [19].

- Zero dimensional (0D): this type of nanomaterial has nano size in all three dimensions. Metal nanoparticles such as gold and silver nanoparticles are a good example of this type of 0D nanoparticles. The majority of these nanoparticle types are spherical and the particle diameters are in the 1–50 nm range.

- One dimensional (1D): this type of nanostructure has one dimension, not at the nanometer range. These include nanowires and nanotubes. These materials are long a (few micrometers in length), while the diameter is just a few nanometers. Nanotubes materials are a good example of this type of nanomaterial.

- Two dimensional (2D): this type of nanomaterial has two dimensions that are not in the nanometer range. These include many different types of nanomaterials and the best example is graphene. The area of this type of nanomaterials may be in the range of a square micrometer, but the thickness remains in the nanoscale size.

- Three dimensional (3D): in this type of nanomaterial all the dimensions are not in the nanometer range. These include bulk materials such as graphite [20, 21].

This review is interested in the most recent type of nanomaterial, which is graphene.

## 2.4 Graphene

The study of graphene is one of the most interesting areas in condensed matter and materials science physics [22]. Moreover, graphene has the potential for many applications in several fields [23, 24]. The plurality of the original research into graphene has focused on its thermal and mechanical properties and analyzed its use in manufacturing applications [25, 26]. A hexagonal ring structure of graphene is formed by a single-atomic layer of $sp^2$ hybridized carbon atoms organized in hexagonal honeycomb structures that are chemically linked to three others with a carbon–carbon bond length of 0.142 nm. Graphite, a 3D layered crystal lattice structure, is formed by stacking parallel 2D graphene sheets. In graphite, weak van der Waals forces hold neighboring graphene sheets together, with a separation distance of 0.335 nm as shown in **Figure 2**. Graphene research has now extended significantly, amidst growing recognition that graphene could have exciting and interesting physical behavior and features such as high stiffness and strength, thermal and electrical conductivity and impermeability to gases (helium, oxygen, nitrogen, etc.). For applications in the nanocomposites field [28, 29], researchers looking at other nanocomposite forms have recently refocused their efforts on graphene nanocomposites. Furthermore, there was pre-existing expertise in graphite exfoliation and the preparation of graphene oxide from graphite oxide. Graphene oxide is related to reduced graphene oxide and pristine graphene by chemical modification [30].

#### *2.4.1 Synthesis of graphene*

The first successful attempts to create a single sheet of graphene using mechanical exfoliation are referred to as the 'scotch tape method'. However, the scotch tape method produces quantities suitable for laboratory research but insufficient volumes for applications as nanocomposites. Much effort is necessary to produce single sheets of graphene. It is necessary to synthesize the monolayer graphene by using procedures like chemical vapor deposition (CVD), epitaxial growth on silicon carbide, molecular beam epitaxy, etc. This approach is known as 'bottom-up' and is not relevant to the

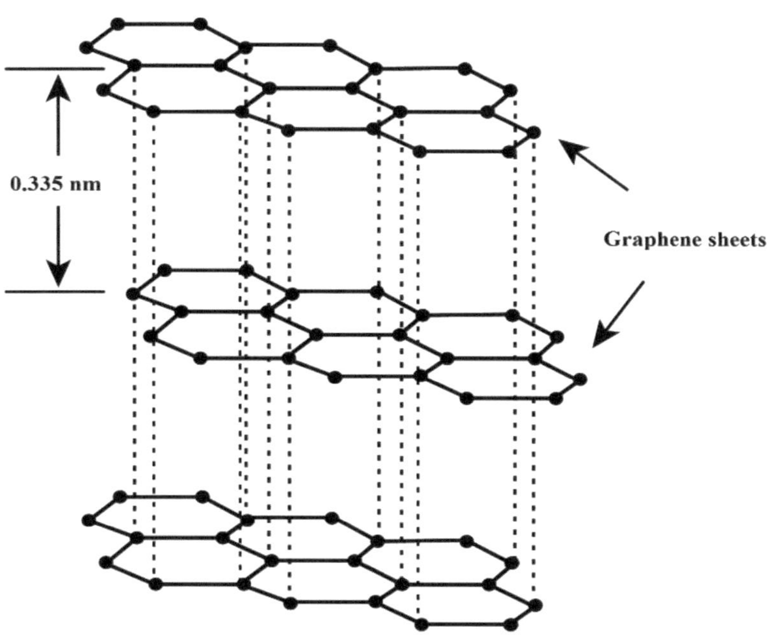


**Figure 2.**
*The layered structure of graphite shows the $sp^2$ hybridized carbon atoms bonded in hexagonal rings [27].*

current project. The other method involves breaking graphite down into graphene sheets using the mechanical cleavage or liquid phase exfoliation known as 'top-down' [31]. In top-down processes, graphene is synthesized in several ways, such as:

- Micromechanical exfoliation of graphite
- Direct sonication of graphite
- Chemical reduction of organically treated graphite oxide (GO)
- Thermal exfoliation/reduction of GO

The graphene used in this project is synthesized using thermal exfoliation and reduction. GO is synthesized by oxidizing the graphite nanoparticles with a mixture of sulfuric acid, sodium nitrate, potassium permanganate and flake graphite. Flake graphite is the most popular source used for graphite oxidation. This certainly creates a form of graphite purified to eliminate heteroatom infection and includes several localized defects to resource the oxidation process. Due to the chemical complexity of flake graphite and the flaws that are present as a result of its natural source, clarifying the specific oxidation mechanism remains difficult. A chemical or a thermal reduction technique can be used to reduce GO levels. Several reducing chemicals, including hydrazine and sodium borohydride, are used to chemically reduce GO sheets. Thermal reduction is a method of removing oxygen functional groups from GO surfaces using heat, this is known as 'the Hummers method' (**Figure 3**) [33, 34].

*2.4.2 Graphene properties and applications*

Graphene is known as one of the most favorable nanomaterials because of its unparalleled group of excellent properties. This provides opportunities for its utilization in a wide variety of applications that can benefit from superior electrical, optical, chemical, thermal and mechanical properties. Electronics, semiconductors, gas absorbers, sensors, solar cells, fuel cells, optic devices and composites can all benefit from graphene. The material's transparency and extremely high conductivity are key features in the most promising graphene applications. Single-layer graphene has a unique electrical structure and capabilities, including a zero bandgap and excellent charge carrier mobility as a result. The material displays transport and conductivity at room temperature. Single-layer graphene also displays room temperature bipolar features, or the quantum Hall effect [35]. Graphene's remarkable characteristics make it ideal for electrical applications. The remarkable electrical properties of graphene, along with its 2D structure and large specific surface area (estimated value of 2630 $m^2g^{-1}$), make it a good gas absorber for next-generation sensors. Due to the existence of $sp^3$ hybridized carbon clusters, the high density of electronegative oxygen atoms bound to carbon, and other imperfections, GO sheets are naturally insulating, with electrical resistance values of roughly 1012/sq. or more. GO can be made electrically conductive through chemical or thermal reduction. The oxygen functions of GO are reduced and the $sp^2$ carbon clusters are restored after heat treatment. This increases electrical conductivity while decreasing electrical resistance [36]. The concentrations of $sp^3$ and $sp^2$ in fully reduced GO are 8% and 80%, respectively. The presence of residual oxygen has been found to significantly inhibit carrier charge (electrons or holes) movement. Furthermore, investigations have revealed that

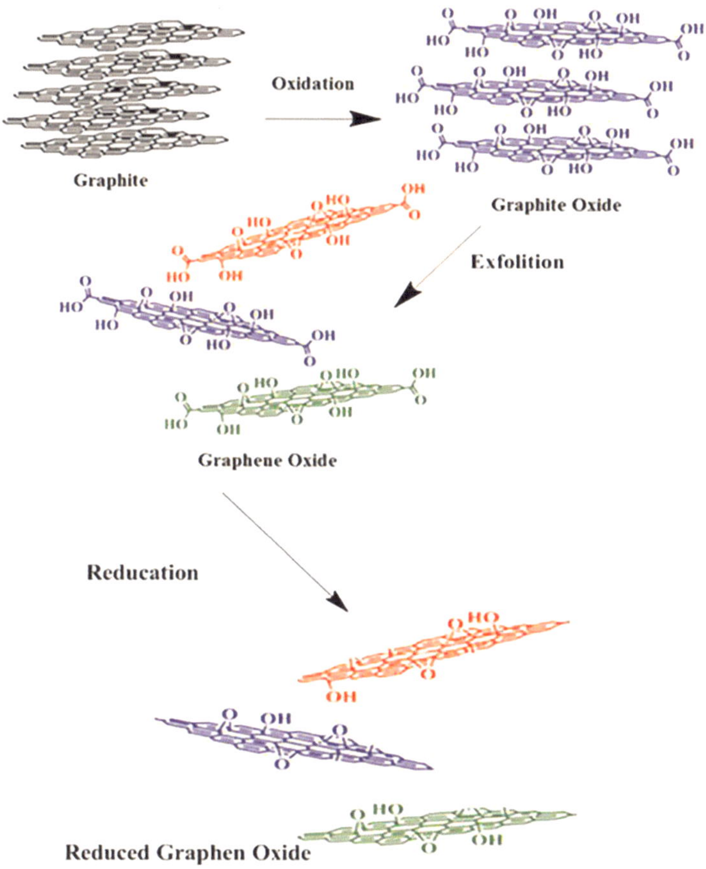


**Figure 3.**
*Synthesis schemes of GO from graphene and rGO from GO [32].*

hopping and tunneling among the $sp^2$ clusters dominate transport in the early stages of reduction, percolation in the latter stages of reduction, and percolation as original $sp^2$ clusters are connected by newly generated tiny domains [37]. Using nanoindentation in AFM, Young's modulus and tensile strength of free-standing monolayer graphene were measured to be 1.0 TPa and 130 GPa, respectively. The mechanical characteristics of GO sheets were investigated using AFM in another study [38], and the effective elastic modulus of monolayer GO (thickness of 0.7 nm) was determined to be 207 $\pm$23 GPa. When oxygen functional groups are attached to graphene, the ideal 2D structure is disrupted, resulting in a considerably weaker GO sheet than pristine graphene. The elastic modulus of GO is greatly influenced by the degree of functionalization and the molecular structure of the functional groups, according to molecular simulation studies [3, 39, 40]. However, graphene exhibits high thermal stability up to 2600 K, dependent on the C/O ratio [41]. Graphene has become a multi-functional reinforcement for polymers due to its unique combination of high electrical, thermal and mechanical properties, which has created new possibilities for developing and enhancing high strength and lightweight polymer composites for the vehicle and space industry, as well as gas barriers for food packaging. To feat graphene's superior properties in such applications, it is necessary to mix it with other materials (such as a polymer) to make stronger and tougher composites known as nanocomposites.

### 2.5 Nanocomposites

Nanocomposites are a combination of two or more distinct materials, in which one is known as the reinforcing phase, which may be in the form of sheets or particles dispersed in the other material known as the matrix phase. The materials are expected to display features, as a result of the combined features of each parent component, that are greater than those of single components. Typically, the host matrix material is improved with just small concentrations of reinforcing materials. For example, if the nanocomposite is designed and fabricated correctly, it may gain reinforcement strength whilst retaining the matrix toughness, thereby exhibiting a combination of desirable properties which are not available in single components [42]. An advantage of nanocomposites, compared to traditional composites, is that such reinforcement should be achieved with the addition of a small percentage of nanoparticulate to the host matrix material. Consequently, the nanocomposites are much lighter weight than traditional composites, but only if the density of the nanoparticle is greater than that of the matrix. Due to that, nanocomposites are a potentially revolutionary alternative to classical composites for many possible applications [43].

### 2.6 Graphene-based polymer nanocomposites

In general, nanocomposites are classified based on the host matrix materials type and type of reinforcement nanoparticles. According to the matrix material type, nanocomposites are classified into three types:

- Polymer matrix-based nanocomposites
- Ceramic matrix-based nanocomposites
- Metal matrix-based nanocomposites [44]

The most common is polymer matrix-based nanocomposites because most polymers display lightweight and high toughness, are easy to process and have high chemical resistance, flexibility and low charge. However, compared to other materials such as metals and ceramics, polymers have comparatively poor mechanical behavior, thermal stability and electrical conductivity. Polymers have as well poor gas barrier and heat resistance properties. The most obvious differentiator of polymers compared with ceramic and metal is weight, due to their lower density. They have low mass atoms of carbon and hydrogen as a backbone, making them suitable for use as lightweight structural components and construction materials. Polymer nanocomposites (PNCs) have been widely studied in industrial and academic fields to identify the unique features of nanosized particles. PNCs have different structures, which can impact the interactions between the polymer and the nanoparticles as filler. Furthermore, composite structures are governed by the type of nanoparticles and the polymer used. The polymers are classified into three different categories: thermoplastic, elastomers and thermosets. These classifications are based on the molecular structure of the polymer. Thermoplastics are often referred to just as plastics, which are linear or branched polymers. They can be molded and remolded many times into different shapes. However, this type of polymer does not easily crystallize on cooling to a solid state, a process that requires the huge organization of the highly coiled and entangled macromolecules present in the liquid state. Thermoplastic polymers cannot fully crystallize because of their inherent structure. The chemical structure of PE and PP polymers supports some degree of crystallization. In many circumstances, crystallization is influenced by experimental factors such as cooling rate and time (in the case of isothermal experiments). When these polymers are cooled from the melt, they are unable to complete crystallization and reach 100% crystallinity. As a result, one component is amorphous and begins to flow at $T_g$ (glass transition temperature), while the crystalline part melts at $T_m$ (melting temperature), creating semi-crystalline polymers. Accordingly, the crystalline phases are characterized by their $T_m$. When a polymer reaches the melting temperature $T_m$, the polymer chains lose their ordered arrangement and move around freely. However, many thermoplastics are completely amorphous, even upon annealing. Amorphous polymers are characterized by the $T_g$, above that, the materials are rubbery or fluid, and below it they are rigid. Semi-crystallinity is a desirable characteristic due to imparting the strength and flexibility of crystalline and amorphous areas, respectively. Consequently, these types of polymers can be rigid with the ability to twist or bend without fracturing. Crystal lamellar is obtained through crystallization from a dilute solution. When crystals are formed from the melt, chain entanglements are quite important. In this case, the solid is more irregular, with polymer chains weaving in and out of crystalline portions. The lamellae are the crystalline part while the amorphous part is the part outside the lamellae. The crystal regions are linked to the amorphous regions by polymer chains. There may be no clear edge limits between those two regions. However, in some polymers, like polyvinyl alcohol, there is notable separation between the crystalline and amorphous regions, though in other polymers, like PE, the structure basically is crystalline with imperfections that are the amorphous regions. The short branches in LDPE interfere with the packing of molecules, so they cannot form a fully ordered structure. The lower density and stiffness make it appropriate for use as films in food packaging and carrier bags [45].

Regarding the structure and properties of the polymer, the nanoparticles and the preparation method, there are three major structural types of composites based on how the filler is dispersed in the polymer. **Figure 4** shows (a) phase-separated micro composites, where the polymer interacts only with the exterior surface of the layered filler, (b) intercalated nanocomposites, in which the layers of filler are sufficiently separated to allow for the polymer to cover each layer and (c) exfoliated nanocomposites, where the layers are separated entirely and dispersed throughout the

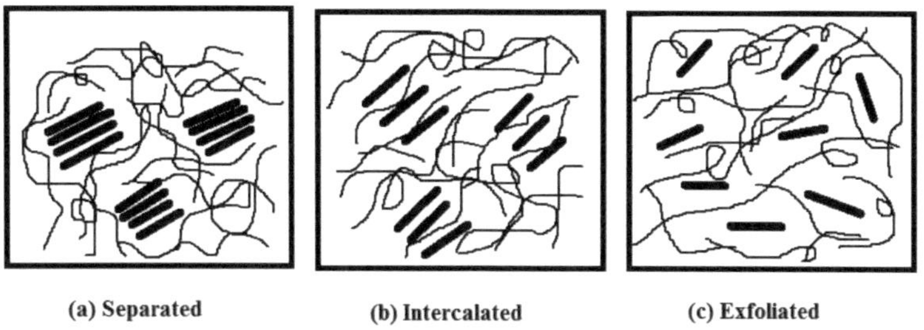


**Figure 4.**
*Filler dispersion in graphene-based nanocomposites: (a) separated, (b) intercalated and (c) exfoliated phases [42].*

polymer phase. Moreover, the PNCs' properties can be enhanced by blending more than one polymer [42].

## 2.7 Preparation methods of polymer nanocomposites

Polymer nanocomposites can be made using chemical, physical and mechanical procedures. One of the major difficulties in polymer nanocomposites fabrication is the good dispersion and distribution of nanofiller into the polymer matrix. Due to the high surface energy of the nanoparticles, they tend to aggregate into micron-sized clusters of filler, which restricts the dispersion and distribution of nanoparticles. Considerable effort has been made to disperse and distribute nanofillers equally into the matrix, regularly assisted by modification of the nanofiller surface, chemical reactions or polymerization reactions which makes them unsuitable for large-scale production [46]. Many preparation methods have been used to prepare graphene nanofiller-reinforced polymer nanocomposites. However, there are three main methods for incorporating the filler into the host polymer matrix [47, 48].

### *2.7.1 In-situ polymerization*

In-situ polymerization methods for polymer nanocomposites fabrication usually include blending the filler in pure monomer, or monomer solution [49], and then the resulting blend is polymerized using polymerization methods such as radiation, heat, initiator diffusion or an organic initiator. The monomer polymerizes between inter-layers, creating either exfoliated or intercalated nanocomposites to improve and develop the dispersion between the two phases. The drawback to this method is that the high-temperature synthesis causes the decomposition of the polymer. Moreover, this process requires the organic modification of the particle surface and employs complex chemical reactions and polymerization reactions [13].

### *2.7.2 Solvent blending*

Solvent blending or solution method is the most commonly used technique for fabricating the PNCs, specifically with higher molecular weight polymers. It involves blending nanoparticles and polymer solutions in a suitable temperature and solvent. Typically, the solution method incorporates three preparation stages: sonicate the nanoparticles in a suitable solvent for the dispersion process, blend it with the polymer solution through simple stirring or shear mixing (at room temperature or higher temperature) and recover the nanocomposite by precipitating or casting the solution mixture as a film. The solution method considerably improves the distribution and dispersion of nanoparticles in the polymer matrix. It offers the advantage of lower viscosity, facilitating regular mixing and good dispersion of the nanoparticles. One of the main drawbacks of this preparation method is the use of large solvents volumes, whose evaporation can impact negatively the environment [11].

### *2.7.3 Melt blending*

Melt compounding is the most common and favorable method used in industry. This method includes blending the nanofillers into the polymer host matrix at a high melting temperature. During the process, the internal shear stress is dissipated in the matrix by viscous drag. That shear stress is applied to break down the nanofiller aggregation and enhance regular and good dispersion of nanofiller in the polymer host matrix. This method is appropriate for present industrial procedures, like extrusion and injection molding. Melt compounding includes melting the polymer powder or pellets to create a viscous solution and the nanofillers are added to the polymer solution at high temperature and high shear. The final form of the components can be produced using compression molding or injection molding. Compared with solution mixing, melt mixing is considered to be more economical and environmentally friendly with no solvent waste and is more effective in the industry due to large production volumes [13, 50]. The polymer matrix nanocomposites can include more than one polymer, which provides opportunities to improve the polymer matrix nanocomposite properties.

### 2.8 The importance of blending polymers

A polymer blend (PB) is a mixture of two or more polymers or copolymers. Blending is a method of obtaining new polymer materials. A mixture of two polymers is referred to as a ‘polymer blend’, ‘poly blends’ or simply ‘blends’. The polymer blend is produced by physical mixing with or without new chemical bonding between the parent components. They are prepared to produce a new material with different physical behavior from the parent polymers. The objective of polymer blending is to achieve sustainable products that either have exclusive properties or lower costs than single polymers. Homogeneous blends are molecular mixed. Heterogeneous blends are thermodynamically immiscible in some concentration ranges. Some blends are prepared for economic reasons, while others are created to improve some properties in the blend. Approximately 10% of all thermoplastics and 75% of all elastomers are poly blends. Only a few commercial blends of two thermoplastics are single-phase blends. All single-phase blends possess negative or slightly positive interaction parameters. They are amorphous blends; their glass temperature varies monotonically with composition. Blends can be compatible but not thermodynamically miscible. Many blends are created from amorphous and (or) semi-crystalline polymers. The majority of these blends are compatible. Blends of two semi-crystalline polymers are rarely used. The components of these blends are usually of a very similar structure. Blending also offers many possibilities for recycling polymer for reuse [51].

### 2.9 Polymer blend methods

The properties of polymer blends depend on the preparation technique. There are many methods used to prepare the polymer blends. The following are the most important and commonly used.

#### *2.9.1 Mechanical-melt mixing*

This is the most important and cheapest method by which to prepare industrial polymer blends. The simplest process for making a polymer blend from thermoplastic is to blend the polymers in a melted condition in suitable devices such as rollers and extruders. Under appropriate conditions, chemical reactions such as chain scissions and cross-linking can take place in the polymer melt. A grafting reaction can also be caused by adding appropriate monomers to polymer melts in extruders [52, 53].

### *2.9.2 Solution cast techniques*

Solution casting is an important method employed to make thin layered films. The solution cast process involves placing the solution of the film component in a suitable common solvent, which is subsequently dried so that the solvent evaporates. The resultant film is then removed from the substrate. A solution casting method is more advantageous than the melting process, as it results in a higher quality film with uniform thickness, high clarity and pure films without residuals and pinholes. It is also possible to produce patterned films [54].

### *2.9.3 Latex blending*

Latex is a colloidal dispersion of a polymer substance in an aqueous medium. Latex blends are prepared by blending two polymers where each polymer is present in the form of polymeric microspheres dispersed in a fluid medium [55]. Blends prepared with this method are expected to have a very high interfacial area. The early emulsion polymerization of rubbers and thermoplastic acrylates provided raw components for latex blending. Latex blends were used either directly as paints, adhesives and sealants or they were pelletized or spray dried [56].

### *2.9.4 Spray or freeze drying*

During spray drying, the fluid of blend materials is transformed into dried particulate form by spraying the fluid into a hot substrate. This is an ideal process used when the end products require precise quality and no remaining moisture content. In freeze drying, the polymers are first heated above the glass transition to form a solution, then the polymer solution is frozen to achieve solid polymer [57].

### *2.9.5 Fine powder mixing*

In this technique, mixtures of polymer powders are mixed at higher temperatures using ball milling. The temperatures used are above the glass transition temperature ($T_g$) of constituent polymers [58].

### *2.9.6 In-situ polymerization*

The polymerization of one polymer is conducted in the presence of another polymer resulting in interpenetrating polymer networks. Polymer electrolytes are prepared using this technique [59]. The preparation method of the blend polymer can impact the miscibility between two or more polymers than their properties.

## 2.10 Properties of polymer blends

Generally, a PB has been prepared to create polymeric materials that can perform under demanding mechanical, chemical, thermal and electrical conditions. They must also be capable of performing in complex atmospheric conditions. All of these factors highlight the necessity of studying the structure, behavior and performance of the PB. The main study for assessing a polymer blend performance is to assess the structure of the blend first since this impacts the material's mechanical, chemical, thermal, flame inhibition, electrical and optical properties. Polymer blends offer excellent advantages such as better processing, superior mechanical (creep, impact, stiffness, strength, modulus and hardness) performance, better heat resistance, lighter weight, gas and water barrier, chemical resistance, optical and electrical properties and low-cost production. PB enables the development and improvement of modified polymers without new polymerization steps [60].

Due to their high molar mass, the mixing entropy of polymers is relatively low and consequently, particularly favorable interactions are necessary to obtain miscible or homogeneous blends on a molecular scale [61]. The overall physical and mechanical behavior depends on the miscibility of the blends, which can be determined by studying the structure of the blend, such as its crystallinity degree, melting and crystallization behavior and phase separation [62–65].

### 2.11 Theory of miscibility

Polymer blends are created when two or more polymers are physically mixed, either in a molten or dissolved state in a suitable solvent. Polymer blends created by the mixing of polymers can be miscible, partially miscible and immiscible. Blends can be also considered compatible or incompatible.

Immiscible blends with separate phases commonly have poor mechanical behavior. Miscible blends with a single phase have various components that are not in a separate phase. That type of polymer blend displays greater mechanical behavior than the parent polymers. However, incompatible (immiscible) blends are more common than compatible (miscible) polymer blends [66, 67].

The most influential factor for obtaining a miscible polymer blend is the low-molecular-weight polymers which have a large combinatorial entropy contribution compared with high-molecular-weight polymers [68].

Whether a polymer blend is miscible, partially miscible or immiscible is determined by the thermodynamics of interaction between the blend components. To obtain spontaneous single-phase blending, the most important factor controlling mixtures of dissimilar components is the Gibbs free energy of mixing ($\Delta G_m$), which should be negative (Eq. (1)).

$$\Delta G_{\mathrm{m}} = \Delta H_{\mathrm{m}} - T\Delta S_{\mathrm{m}} \leq 0 \tag{1}$$

where $\Delta H_{\mathrm{m}}$ and $\Delta S_{\mathrm{m}}$ are the enthalpy and entropy of mixing, respectively, and $T$ is temperature. This is, however, a necessity but not a sufficient condition. Furthermore, a second condition should also be fulfilled for each blend composite to attain a single-phase binationry polymer mixture (Eq. (2)).

$$\left(\frac{\partial^2 G_{\mathrm{m}}}{\partial \varphi_i^2}\right) > 0 \tag{2}$$

where $\varphi_i$ is a volume fraction of the component. If $\Delta G_{\mathrm{m}}$ is negative and Eq. (2) is not fulfilled, the polymer blend will separate into two phases. **Figure 4** shows a generic phase diagram for polymer blend systems. The spinodal curve is related to the condition (Eq. (3)).

$$\left(\frac{\partial^2 G_{\mathrm{m}}}{\partial \varphi_i^2}\right)_{T,P} = 0 \tag{3}$$

Within these curves, the polymer mixture is unstable and will undergo spinodal decomposition. External to the spinodal curve lay the stable and metastable regions. The transition between these two regions is the binodal curve, which is where $\Delta G_m = 0$. Blends in the metastable region will spontaneously nucleate due to composition fluctuations and separate into continuous and dispersed phases. Blends in the stable region will undergo spontaneous mixing and exhibit a single, homogenous phase [69, 70].

For low molecular weight materials, an increased temperature mostly leads to greater miscibility as the $T\Delta S_m$ term increases, thus driving $\Delta G_m$ to more negative values. For higher molecular weight components, the $T\Delta S_m$ term is small and other factors (such as non-combinatorial entropy contributions and temperature dependant $\Delta H_m$ values) can dominate and lead to the reverse behavior, namely, decreasing miscibility with increasing temperature.

Solvent blends that are borderline in miscibility normally show upper critical solution temperatures (UCST) and polymer-polymer mixtures normally show lower critical solution temperatures (LCST) [71, 72]. This behavior is shown in **Figure 5**.

The simplest theory to calculate $\Delta Gm$ for a component of two polymers blends is the Flory-Huggins expression. That theory for the free energy of mixing of polymer-solvent systems has been extended to include polymer-polymer mixtures (Eq. (4)).

$$\frac{\Delta G_m}{VRT} = \frac{\varphi_1}{\upsilon_1 N_1} \ln \varphi_1 + \frac{\varphi_2}{\upsilon_{2\,N_2}} \ln \varphi_2 + \varphi_1 \varphi_2 \frac{\chi}{\upsilon} \qquad (4)$$

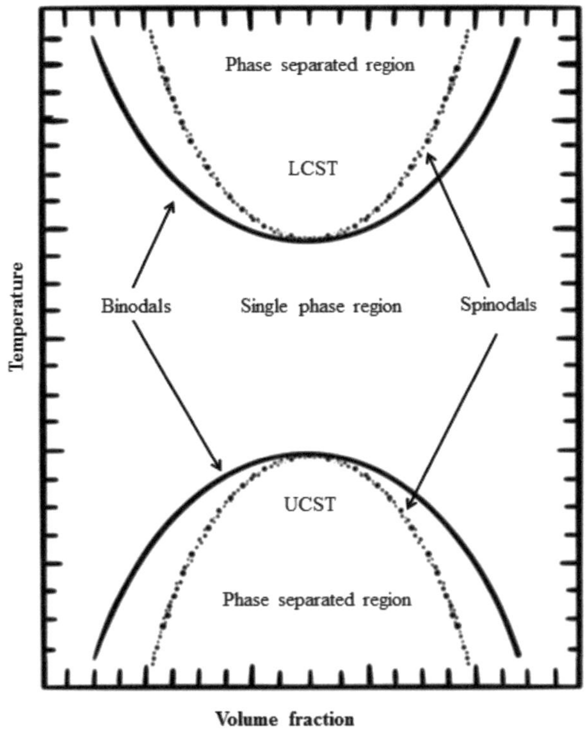


**Figure 5.**
*Phase diagram showing LCST and UCST behavior for polymer blends [73].*

where $V$ is the total volume of specimen, $T$ is absolute temperature, $\upsilon_i$ is the molar component volume of component $i$, $N_i$ is the degree of component $i$ polymerization, $\varphi_i$ is the component $i$ volume fraction, $\upsilon$ is a reference volume and $\chi$ is the Flory-Huggins interaction parameter. The first two terms on the right-hand side of Eq. (4) account for the $T\Delta S_m$ term in Eq. (1) and are the entropic contribution of each component to mixing. The final term on the right describes the enthalpic mixing contribution to $\Delta G_m$.

For large-size components ($N_i > 1000$), the respective entropic term becomes negligible. A critical spontaneous mixing interaction parameter $\chi_c$ can be defined when $\Delta G_m = 0$ and $\frac{\partial^2 G_m}{\partial \varphi_i^2} = 0$ which leads to Eq. (5).

$$\chi_c = \frac{\upsilon}{2}\left(\frac{1}{\sqrt{\upsilon_1 N_1}} + \frac{1}{\sqrt{\upsilon_2 N_2}}\right)^2 \quad (5)$$

Eq. (5) can be simplified further when both components have equal degrees of polymerization ($N_1 = N_2$) and molecular volumes ($\upsilon_1 = \upsilon_2$), yielding (Eq. (6)).

$$\chi_c N = 2 \quad (6)$$

Comparing the $\chi$ $N$ values for different blend systems to this critical $\chi_c$ $N$ value allows for a first-order determination of whether a polymer blend is miscible, partially miscible or immiscible leading to a more complex two-phase system [74].

### 2.12 The classification of the polymer blend

Polymer blends are classified into the following types: miscible polymer blends, partially miscible polymer blends and immiscible polymer blends [73].

#### *2.12.1 Miscible polymer blends*

Compatible blends are characterized by a $\chi N < 2$ and can create a single phase or homogeneous phase due to short chain lengths or suitable enthalpies $\Delta H_m$ of blending. Complementary intermolecular forces between side groups, such as acid–base interactions, hydrogen bonds, dipoles, ionic groups and π-orbital complexes, frequently induce negative $\Delta H_m$ values. Polystyrene and poly (2, 6-dimethylphenylene oxide) (PS/PPO), for example, are miscible due to the interaction of styrene groups. These blends have improved physical properties and a single-glass transition temperature, indicating that just one phase exists [75].

#### *2.12.2 Partially miscible polymer blends*

Binary blends of polymer that have a $\chi N$ value of ≈2 can be classified as partially miscible. These blends show two distinct phases with a quite broad interface region separating them. Moreover, there are sufficient concentrations of minority components in both phases to modify the bulk properties. The mixture of polystyrene and acrylonitrile butadiene styrene (PS/ABS) belongs to this type. Blends show two glass transition temperatures ($T_g$), with the $T_g$ of each component shifting slightly toward the other compared to pure polymer transitions. The separated phases limit the deformation mechanisms of the blend, which promotes irreversible micro crazing [76]. This blending regime can be referred to as 'compatible blends'.

### *2.12.3 Immiscible polymer blends*

Polymer blends with $\chi N > 2$ are referred to as immiscible (incompatible) polymer blends. The notable feature of this blend is two distinct phases separated by a sharp interface. The interfaces have minor interaction between the two phases and as a result show very poor cohesion. Therefore, the physical properties of immiscible blends are almost poorer than either of the parent polymers alone. When a blend solidifies, the minor, dispersed phase, thermodynamically favors specific geometries depending on the blend composition [77].

## 2.13 Compatibilization of immiscible blends

Due to the positive Gibbs energy of mixing, most polymers are immiscible, resulting in severe phase separation, poor adhesion at surfaces and decreased final characteristics. Many strategies have recently been employed to generate and improve interactions in immiscible polymer blends' interfacial regions. Using proper compatibilizers is the simplest and most successful way to make immiscible polymers compatible. Compatibilization is the process of adding appropriate copolymers to an immiscible polymer blend to increase its stability. In most polymer mixes, the phase morphologies are unstable, resulting in poor mechanical characteristics. Compatibilizing the system leads to a stable and better-mixed phase morphology by generating interaction between the two immiscible polymers [78]. There are many different types of compatibilizers such as graft or block copolymers [79] and nanofillers such as clay or layered silicates [80], carbon nanotubes [81] and graphene oxide (GO) or its derivatives [82, 83].

### *2.13.1 Compatibilization using Ziegler-Natta catalyzed copolymers*

Ziegler-Natta catalysts are the most common catalysts used in polymerization manufacturing for the production of PE and PP [84]. Based on solubility, the Ziegler-Natta catalyst can have categorized as either:

- Heterogeneous catalysts: These catalysts are usually in combination compounds, such as poly (ethylene-co-glycidyl methacrylate) (PE-co-GMA) for PE and PP blend co-catalyst [85, 86].

- Homogeneous catalysts: random copolymers of ethylene-α-olefin and metallocene linear low-density polyethylene (PE) are obtained using metallocene catalyst [87, 88].

As compatibilizing agents, graft or block copolymers are frequently utilized. Two polymers in an immiscible blend are used in the copolymers used for that purpose. The appropriate components of the copolymer can interact with the blend's two phases to help maintain the phase morphology's stability [89, 90]. The increase in stability is attributable to a reduction in the phase separation size of the polymers in the blend. The lower interfacial tension caused by the accumulation of block copolymers at the interfaces of the two polymers results in a smaller size. In the melt phase, this allows the immiscible mixes to break up into smaller particles. Because the interfacial tension is now significantly lower, these phase-separated particles will be less likely to proliferate. This stabilizes the polymer blend, allowing it to be used in a product [91]. Ethylene/propylene copolymers are an example of this. Polypropylene and low-density polyethylene blends benefit from compatibilizing polymers like these. In this situation, a copolymer with a longer ethylene series is favored. This is because co-crystallization is a factor in this situation, and the longer ethylene series will maintain some crystallinity [92, 93]. Due to their low molecular weights compared to that of the bulk polymer, copolymers often bring minor benefits to the polymer blend's strength and stiffness. Furthermore, copolymers with certain structures are frequently difficult to synthesize, making them expensive to produce. As a result, another effective and low-cost compatibilization technique is required [94–96].

#### *2.13.2 Graphene compatibilization*

Inorganic nanofillers are promising for compatibilizing immiscible polymer blends because of their good performance and inexpensive cost. GO is a result of oxidative exfoliation of natural graphite that consists of multilayers of sp2-hybridized carbon atoms on the basal plane and edges with a variety of carboxyl, hydroxyl and epoxy functional groups [85, 97–100]. The polar groups of GO layers can create hydrogen bonds with polar polymers [101]. Some polymers' π-π stacking phenomena between GO and aromatic rings could be used to improve interfacial interaction, resulting in improved compatibility and mechanical strength [102]. Due to their high modulus, GO and its derivatives may not only operate as a compatibilizer in polymer blends but also as reinforcing fillers, making GO and its derivatives superior to a traditional copolymer compatibilizer. The processing technique allows for the required improvement in characteristics and structure in addition to the surface modification of the filler and the use of a suitable compatibilizer. Because nanofillers have a strong tendency to clump and agglomerate due to their high surface energy, the processing system's efficiency in dispersing the fillers becomes crucial [103].

### 2.14 Properties of graphene/based polymer nanocomposites

In general, nanocomposites need to be thermally stable for many applications, such as in the aerospace industry. However, most polymer nanocomposites suffer from low-temperature degradation, which limits their use in many possible applications. Graphene, which is one sheet of graphite, has unique features such as high conductivity, strength and thermal stability. This exceptional material can be incorporated into the polymer composites as nanofillers to enhance properties.

As a result of the fact that graphene/polymer nanocomposites are a relatively recent development, the literature is still in its early stages but is developing rapidly. Already surprising developments and enhancements in mechanical, electrical and thermal properties and water and gas barriers of these materials have been achieved at very low concentrations of graphene nanoparticles in the polymer host matrix.

#### *2.14.1 The mechanical properties*

Graphene is considered a strong material that offers the possibility to make refinements in the mechanical behaviors of polymeric materials at low concentrations, in particular enhancing the tensile strength and Young's modulus as shown in **Figure 6**.

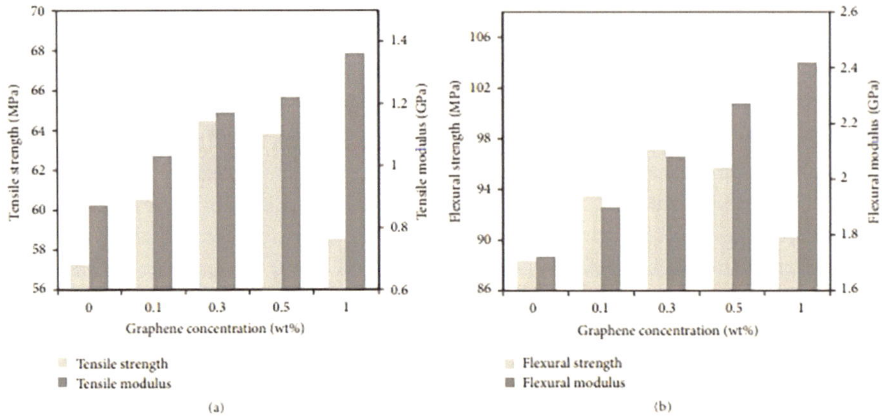


**Figure 6.**
*Mechanical properties of epoxy/graphene nanocomposites: (a) tensile properties and (b) flexural properties [104].*

However, the mechanical properties of PNCs depend on the dispersion and distribution of graphene flakes into the polymer matrix and the degree of interfacial bonding between the nanofiller and the matrix. Generally, to develop and improve the mechanical behavior of graphene/polymer nanocomposites, GO is typically used due to its outstanding mechanical behavior and the presence of chemical groups to assist in strong interfacial interaction [105]. Although pure graphene is incompatible with polymers and will aggregate, through layer-by-layer stacking, GO, which contains hydroxyl and epoxy groups on the plane of the sheet and carbonyl and carboxyl groups at the edges, interacts more strongly with polymers. Its surface is comparatively easy to modify through the presence of amines, esters, aromatics and isocyanate functionalities that stabilize dispersions; thereby facilitating treating the composite [106]. However, the graphene nanoparticles could improve another important feature in polymer/graphene nanocomposites which is the hardness. **Figure** 7 shows a significant improvement in hardness observed with the addition of graphene. The neat

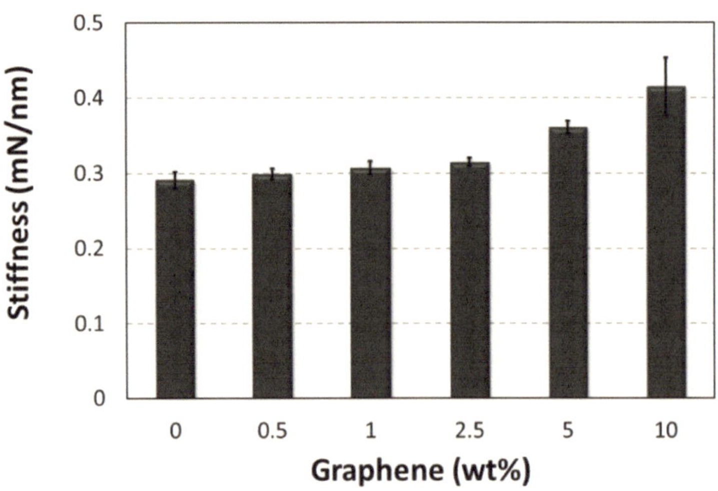


**Figure 7.**
*The effect of graphene content on the hardness of the graphene/epoxy nanocomposites [107].*

epoxy samples exhibit a hardness of 26.56 kgf/mm$^2$, which increases to 28.90 (8.83%), 31.59 (19%), 33.37 (26%), 35.02 (32%) and 39.89 (50%) with the addition of 0.5 1, 2.5, 5 and 10 wt % graphene nanoparticles, respectively. This increment in hardness is due to a good dispersion and interfacial bonding between graphene and the epoxy polymer matrix [107].

### *2.14.2 Electrical properties*

Electrical behavior is one of the most interesting features of graphene/polymer nanocomposites, enabling them to be employed in electronics applications because of their low electrical resistance and high electrical conductivity. When used as filler, graphene might raise the material conductivity of an insulator polymer, such as poly (vinyl alcohol) in **Table 1**. To make the nanocomposites conductive, the percentage of the conducting filler should be above the electrical percolation threshold where a conductive network of nanoparticles filler is formed. However, GO is an insulator and not an ideal filler for producing electrically conducting composites. Another method involves GO surface modification, reduction to recover, at least partially, the electrical and thermal conductivity through restoring the graphitic network of the sp$^2$ hybridized carbon by reducing the carbon–oxygen function to have reduced graphene oxide or even pristine graphene [104, 109–111].

### *2.14.3 Thermal stability*

When graphene/polymer nanocomposites are prepared through the solvent method, a well-dispersed system is created maximizing the graphene surface area, which will contribute to superb material thermal stability [112, 113]. Covalent modifications can be used to improve the graphene dispersion and distribution into the polymer matrix to attain the most significant effect. The modification could also change the microstructure of graphene, resulting in graphene with a high carbon-to-oxygen atom ratio (C/O ratio). This may increase thermal stability, through a decrease in the oxygen functional group content and an increase in the C/O ratio as shown in **Figure 8** and **Table 2** [115, 116]. In general, a tiny loaded amount of graphene can notably enhance the thermal stability of polymer materials [117, 118].

### *2.14.4 Thermal conductivity*

All electronic units produce excessive heat and thus demand thermal management to prevent premature failure. Thermal management is crucial for the efficiency of

| Composites | Length (m) | Width (m) | Area (m$^2$) | Resistance (Ω) | Resistivity (Ωm) | Conductivity (Ωm$^{-1}$) |
|---|---|---|---|---|---|---|
| PVA | 0.01 | 0.008 | 0.00008 | 2398 | 19 | 0.05 |
| PVA/GO | 0.01 | 0.008 | 0.00008 | 1644 | 13 | 0.073 |
| PVA/RGO | 0.01 | 0.008 | 0.00008 | 199 | 1.6 | 0.625 |

*PVA, Poly(vinyl alcohol); PVA/GO, poly(vinyl alcohol)/graphene oxide; PVA/RGO, poly(vinyl alcohol)-reduced graphene oxide.*

**Table 1.**
*Resistance, resistivity and conductivity of the nanocomposites [108].*

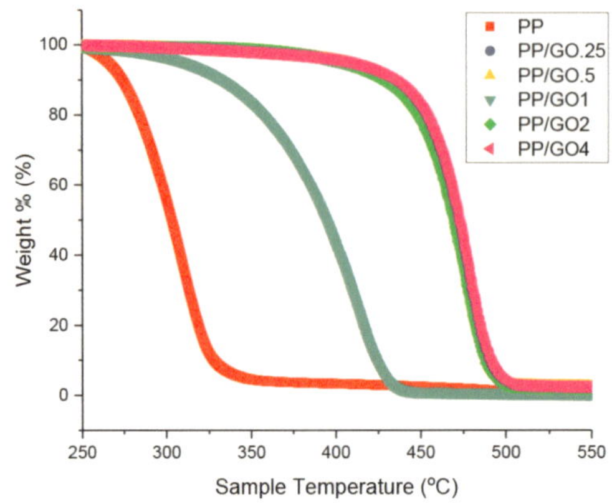


**Figure 8.**
*Thermal degradation temperatures PP/GO nanocomposites [114].*

| Specimens ID | GO (wt%) | $T_{onset}$ (°C) | $T_{50\%}$ (°C) | $T_{95\%}$ (°C) |
|---|---|---|---|---|
| PP | 0.00 | 276 ± 3 | 303 ± 3 | 344 ± 3 |
| PP/GO.25 | 0.25 | 440 ± 3 | 472 ± 3 | 494 ± 3 |
| PP/GO.5 | 0.50 | 452 ± 3 | 473 ± 3 | 496 ± 3 |
| PP/GO1 | 1.00 | 409 ± 3 | 443 ± 3 | 481 ± 3 |
| PP/GO2 | 2.00 | 445 ± 3 | 470 ± 3 | 494 ± 3 |
| PP/GO4 | 4.00 | 453 ± 3 | 473 ± 3 | 497 ± 3 |

*PP, polypropylene; PP/GO, polypropylene/graphene oxide.*

**Table 2.**
*Thermal degradation temperatures PP/GO nanocomposites [114].*

advanced integrated circuits (ICs) and high-frequency high-power density communication devices. Recently, the use of high-conductivity materials is suggested for electronic cooling and for improving the heat dissipated from chips. The cost of high-conductivity materials is of major concern. Therefore, there is a real need for low-cost high-thermal conductivity materials and efficient design to integrate these materials into electronic devices. Graphene has drawn tremendous attention for heat dissipation due to its extraordinarily high in-plane thermal conductivity (2000–4000 $Wm^{-1}K^{-1}$) compared to copper (400 $Wm^{-1}K^{-1}$). The thermal conductivity of graphene has become an important research topic and is attracting tremendous interest in the area of thermoelectric waste heat recovery. The thermal conductivity of graphene is related to its low mass and the strong bond of carbon atoms [119].

### *2.14.5 Gas barrier*

Graphene and its derivatives have been considered promising nanoscale fillers in the gas barrier application of polymer nanocomposites (PNCs). The breakthrough of gas into polymer films has limited their performance. The barrier properties of polymers can be greatly improved by loading impermeable lamellar fillers with a high aspect ratio to change the diffusion path of gas-penetrant molecules, such as graphene.

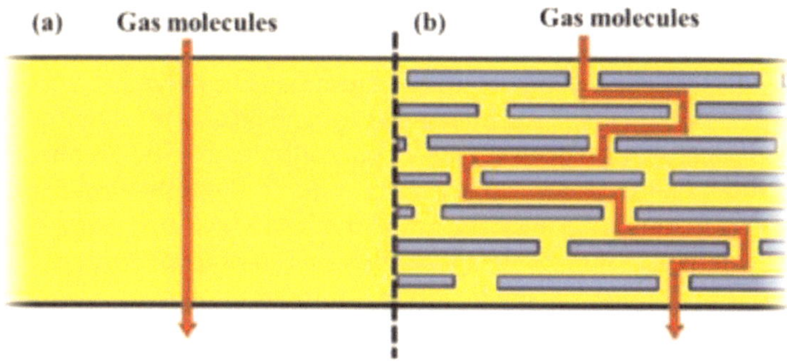


**Figure 9.**
*Illustration of the 'tortuous pathway' (a) In a polymer matrix before adding graphene nanoparticles (b) In nanocomposite after adding the graphene nanoparticles [120].*

As a result of the nanofillers, diffusing molecules take longer and more tortuous paths to pass through the nanocomposite film, resulting in a considerable decrease in permeability as shown in **Figure 9**. The gas barrier performance of PNCs is determined by mainly three factors: filler properties (resistance to gas diffusion, aspect ratio and volume fraction), the intrinsic barrier property of the polymer matrix, and the 'quality' of dispersion (agglomeration/specific interface, free volume generated by mediocre interface management and the texture/orientation of filler platelets). The levels of exfoliation of the layered nanofillers in the polymer matrix are crucial to the successful development of PNCs [121].

## 3. Conclusions

Since its inception, graphene has drawn worldwide interest due to its high quality and amount of work. The technology and applications of graphene-reinforced polymer matrix nanocomposites are showing significant development. However, various challenges in concerns of synthesis methods, costs, quantity qualities and applications must be addressed and resolved to fully realize the potential of graphene nanocomposites. For illustration, if we consider physical synthesis methods for graphene such as sonication, exfoliation and cleaving, the end product (graphene) can have a reduced aspect ratio, which can drastically degrade the reinforcement, and binding interactions, and thermal and electrical properties of both graphene and nanocomposites. Both graphene and its derivatives have proved their promise as prospective choices as reinforcements for high-performance nanocomposites, as shown in this review. The effects of reduced loadings of graphene and its derivatives, which result in high levels of strength and stiffness and superior mechanical properties, have been described in many studies. The high dispersion quality of graphene and its derivatives in various host matrices has also been discussed in many types of research. As previously stated, several challenges must be overcome in achieving large-scale, defect-free graphene exfoliation with high quality and good characteristics. Since graphene tops the charts with its exceptional properties, the graphene-based product highlights itself in various applications. Dispersion and distribution of graphene into polyolefin are very important to develop and improve the electrical and thermal conductivity, mechanical properties and gas barriers of the polymers. To attain most of these characteristics, it is necessary to have exfoliation, an interconnected graphene network.

However, dispersions and distribution of graphene in polymers are challenging. Partially get good dispersion by sonicating graphene into a co-solvent with the polymer before co-precipitation or drying to a film, or by scattering graphene in a monomer and polymerizing in situ to make a composite. In the right circumstances, solvent-processed composites lock the graphene into a well-dispersed state in the solvent; nevertheless, to improve dispersibility and interaction with the polymer matrix, the graphene must usually be modified. In-situ polymerization can produce better outcomes, which could be because graphene is involved in the polymerization process, or because polymer chains are grafted onto graphene sheets.

## Acknowledgements

This chapter was supported by IntechOpen (www.intechopen.com).

## Conflict of interest

There are no known conflicts of interest associated with this publication and there has been no financial support for this work that could have influenced its outcome.

## Author details

Haia Aldosari
Department of Physics and Astronomy, Shaqra University, Shaqra, Saudi Arabia

*Address all correspondence to: haldossari@su.edu.sa

## References

[1] Huang C. Carbon black filled conducting polymers and polymer blends. Advances in Polymer Technology. 2002;**21**:299-313

[2] Moniruzzaman M, Winey K. Polymer nanocomposites containing carbon nanotubes. Macromolecules. 2006;**39**: 5194-5205

[3] Suk JW, Piner RD, An J, Ruoff RS. Mechanical properties of monolayer graphene oxide. ACS Nano. 2010;**4**: 6557-6564

[4] Snook GA, Kao P, Best AS. Conducting-polymer-based supercapacitor devices and electrodes. Journal of Power Sources. 2011;**196**:1-12

[5] Wan X, Huang Y, Chen Y. Focusing on energy and optoelectronic applications: A journey for graphene and graphene oxide at large scale. Accounts of Chemical Research. 2012;**45**:598-607

[6] Yin Z, Wu S, Zhou X, Huang X, Zhang Q, Boey F, et al. Electrochemical deposition of ZnO nanorods on transparent reduced graphene oxide electrodes for hybrid solar cells. Small. 2010;**6**:307-312

[7] Yin ZY, Sun SY, Salim T, Wu SX, Huang XA, He QY, et al. Organic photovoltaic devices using highly flexible reduced graphene oxide films as transparent electrodes. ACS Nano. 2010; **4**:5263-5268

[8] Yavari F, Koratkar N. Graphene-based chemical sensors. Journal of Physical Chemistry Letters. 2012;**3**: 1746-1753

[9] Mishra SK, Tripathi SN, Choudhary V, Gupta B. SPR based fibre optic ammonia gas sensor utilizing nanocomposite film of PMMA/reduced graphene oxide prepared by in situ polymerization. Sensors and Actuators, B: Chemical. 2014;**199**:190-200

[10] Cromer BM, Scheel S, Luinstra GA, Coughlin EB, Lesser AJ. In-situ polymerization of isotactic polypropylene-nanographite nanocomposites. Polymer (United Kingdom). 2015;**80**:275-281

[11] Chee WK, Lim HN, Huang NM, Harrison I. Nanocomposites of graphene/polymers: A review. RSC Advances. 2015;**5**:68014-68051

[12] Park S, An J, Potts JR, Velamakanni A, Murali S, Ruoff S. Hydrazine-reduction of graphite- and graphene oxide. Carbon. 2011;**49**:3019-3023

[13] Tripathi SN, Rao GS, Mathur AB, Jasra R. Polyolefin/graphene nanocomposites: A review, RSC Advances. 2017;**7**:23615-23632

[14] Barkoula NM, Alcock B, Cabrera NO, Peijs T. Fatigue properties of highly oriented polypropylene tapes and all-polypropylene composites. Polymers and Polymer Composites. 2008;**16**:101-113

[15] Dowling A, Clift R, Grobert N, Hutton D, Oliver R, O'neill O. Nanoscience and nanotechnologies: Opportunities and uncertainties. London The Royal Society The Royal Academy of Engineering Report. 2004;**46**:618-618

[16] Kittelson DB. Recent measurements of nanoparticle emission from engines current research on diesel exhaust particles. Japan Association of Aerosol Science and Technology. 2001;**9**

[17] Borm P, Robbins D, Haubold S, Kuhlbusch T, Fissan H, Donaldson K, et al.

The potential risks of nanomaterials: A review carried out for ECETOC (review). Particle and Fibre Toxicology. 2006:3-11

[18] Cristina B, Ivan I, Pacheco B, Kevin R. Nanomaterials and nanoparticles: Sources and toxicity. Biointerphases. 2007;**2**:17-172

[19] Teo B, Sun X. Classification and representations of low-dimensional nanomaterials: Terms and symbols. Journal of Cluster Science. 2007;**18**: 346-357

[20] Churg A. Interactions of exogeneous or evoked agents and particles: The role of reactive oxygen species. Free Radical Biology and Medicine. 2003;**34**: 1230-1235

[21] Donaldson K, Stone V, Borm A, Jimenez A, Gilmour S, Schins F, et al. Oxidative stress and calcium signaling in the adverse effects of environmental particles. Free Radical Biology and Medicine. 2003;**34**:1369-1382

[22] Geim AK, Novoselov KS. The rise of graphene. Nature Materials. 2007;**6**: 183-191

[23] Geim AK. Nobel lecture: Random walk to graphene. Reviews of Modern Physics. 2011;**83**:851-862

[24] Novoselov KS. Nobel lecture: Graphene: Materials in the flatland. Reviews of Modern Physics. 2011;**83**: 837-849

[25] Avouris P. Graphene: Electronic and photonic properties and devices. Nano Letters;**10**(1010):4285-4294

[26] Schwierz F. Graphene transistors. Nature Nanotechnology. 2010;**5**:487-496

[27] Sengupta R, Bhattacharya M, Bandyopadhyay S, Bhowmick AK. A review on the mechanical and electrical properties of graphite and modified graphite reinforced polymer composites. Progress in Polymer Science (Oxford). 2011;**36**:638-670

[28] Kim H, Abdala A, Macosko CW. Graphene/polymer nanocomposites. Macromolecules. 2010;**43**:6515-6530

[29] Singh V, Joung D, Zhai L, Das S, Khondaker SI, Seal S. Graphene based materials: Past, present and future. Progress in Materials Science. 2011;**56**: 1178-1271

[30] Verdejo R, Bernal MM, Romasanta L, Lopez-Manchado M. Graphene filled polymer nanocomposites. Journal of Materials Chemistry. 2011;**21**: 3301-3310

[31] Young R, Kinloch I, Gong L, Novoselov KS. The mechanics of graphene nanocomposites: A review. Composites Science and Technology. 2012;**72**:1459-1476

[32] Sengupta J, Hussain C. Graphene-based field-effect transistor biosensors for the rapid detection and analysis of viruses: A perspective in view of COVID-19. Carbon Trends. 2021;**2**: 100011

[33] Gao W. The chemistry of graphene oxide. Graphene Oxide: Reduction Recipes, Spectroscopy, and Applications. 2015:61-95

[34] Layek RK, Samanta S, Nandi AK. The physical properties of sulfonated graphene/poly(vinyl alcohol) composites. Carbon. 2012;**50**:815-827

[35] Choi W, Lahiri I, Seelaboyina R, Kang Y. Synthesis of graphene and its applications: A review. Critical Reviews in Solid State and Materials Sciences. 2010;**35**:52-71

[36] Wang S, Geng Y, Zheng Q, Kim JK. Fabrication of highly conducting and transparent graphene films. Carbon. 2010;**48**:1815-1823

[37] Mattevi C, Eda G, Agnoli S, Miller S, Mkhoyan KA, Celik O, et al. Evolution of electrical, chemical, and structural properties of transparent and conducting chemically derived graphene thin films. Advanced Functional Materials. 2009;**19**: 2577-2583

[38] Wang C, Frogley MD, Cinque G, Liu L, Barber AH. Deformation and failure mechanisms in graphene oxide paper using in situ nanomechanical tensile testing. Carbon. 2013;**63**:471-477

[39] Zheng Q, Li Z, Jialin Y, Kim J. Molecular dynamics study of the effect of chemical functionalization on the elastic properties of graphene sheets. Journal of Nanoscience and Nanotechnology. 2010;**10**:7070-7074

[40] Lee C, Wei X, Kysar J, Hone J. Measurement of the elastic properties and intrinsic strength of monolayer graphene. Science. 2008;**321**:385-388

[41] Kim K, Regan W, Geng B, Alemán B, Kessler BM, Wang F, et al. High-temperature stability of suspended single-layer graphene. Physica Status Solidi—Rapid Research Letters. 2010;**4**: 302-304

[42] Henrique P, Camargo C, Satyanarayana K, Wypych F. Nanocomposites: Synthesis, structure, properties and new application opportunities. Materials Research. 2009; **12**:1-39

[43] Kerstin M, Elodie B, Marcos L, Maria J, Yolanda E, José L, et al. Review on the processing and properties of polymer nanocomposites and nanocoatings and their applications in the packaging, automotive and solar energy fields. Nanomaterials. 2017:7-47

[44] Young R, Lovell P. Introduction to Polymers. 3rd ed. London, UK: CRC Press; 2011

[45] In-Yup J, Jong-Beom B. Nanocomposites derived from polymers and inorganic nanoparticles. Materials. 2010;**3**:3654-3674

[46] Deng H, Lin L, Ji M, Zhang S, Yang M, Fu Q. Progress on the morphological control of the conductive network in conductive polymer composites and the use as electroactive multifunctional materials. Progress in Polymer Science. 2014;**39**:627-655

[47] Kuilla T, Bhadra S, Yao D, Kim N, Bose S, Lee J. Recent advances in graphene-based polymer composites. Progress in Polymer Science. 2010;**35**: 1350-1375

[48] Sadasivuni K, Ponnamma D, Kim J, Thomas S. Graphene-based polymer nanocomposites in electronics. Graphene-Based Polymer Nanocomposites in Electronics. 2015: 1-382

[49] Potts J, Dreyer D, Bielawski C, Ruoff S. Graphene-based polymer nanocomposites. Polymer. 2011;**52**:5-25

[50] Nguyen Q, Baird D. Preparation of polymer–clay nanocomposites and their properties. Advances in Polymer Technology. 2006;**25**:270-285

[51] Kalaitzidou K, Fukushima H, Drzal L. A new compounding method for exfoliated graphite-polypropylene nanocomposites with enhanced flexural properties and lower percolation threshold. Composites Science and Technology. 2007;**67**:2045-2051

[52] Shemouratov Y, Prokhorov K, Sagitova E, Nikolaeva G, Pashinin P, Lebedev Y, et al. Raman study of polyethylene-polypropylene blends. Laser Physics. 2009;**19**:2179-2183

[53] Braun D, Cherdron H, Rehahn M, Ritter H, Voit B. Polymer Synthesis: Theory and Practice. 4th ed. Berlin Heidelberg: Springer-Verlag; 2005. p. 12

[54] Samperi F, Battiato S, Recca G, Puglisi C, Mendichi P. Reactive melt mixing of PC/PEN blend, structural characterization of reaction products. Polymer. 2015;**74**:108-123

[55] Topham P, Howse J, Fernyhough C, Ryan A. The performance of poly (styrene)-block-poly(2-vinyl pyridine)-block-poly(styrene) triblock copolymers as pH-driven actuators. Soft Matter. 2007;**3**:1506-1512

[56] Feng J, Winnik M, Shivers R, Clubb B. Polymer blend latex films: Morphology and transparency. Macromolecules. 1995;**28**:7671-7682

[57] Song P, Cao Z, Cai Y, Zhao L, Fang Z, Fu S. Fabrication of exfoliated graphene-based polypropylene nanocomposites with enhanced mechanical and thermal properties. Polymer. 2011;**52**:4001-4010

[58] Paudel A, Worku Z, Meeus J, Guns S, Van Den Mooter G. Manufacturing of solid dispersions of poorly water soluble drugs by spray drying: Formulation and process considerations. International Journal of Pharmaceutics. 2013;**453**:253-284

[59] Tiwari R, Srivastava B, Tiwari G, Rai A. Extended release promethazine HCl using acrylic polymers by freeze-drying and spray-drying techniques: Formulation considerations. Brazilian Journal of Pharmaceutical Sciences. 2009;**45**:829-840

[60] Koch C. Top-down synthesis of nanostructured materials: Mechanical and thermal processing methods. Reviews on Advanced Materials Science. 2003;**5**:91-99

[61] Ogawa T. Poly(silmethylene )-based polymer blends. I. *In situ* polymerization in silicon-based polymers. 1997:399-405

[62] Shanks R, Li J, Yu L. Polypropylene—polyethylene blend morphology controlled by time-temperature—miscibility. Materials Science. 2000;**41**:2133-2139

[63] Walsh D, Rostami S. The miscibility of high polymers: The role of specific interactions. Advances in Polymer Science. 1984:120-163

[64] Scobbo J, Goettler L. Applications of polymer alloys and blends. In: Utracki LA, editor. Polymer Blends Handbook. Kluwer Academic Publishers; 2003. pp. 951-976

[65] Yu L, Dean K, Li L. Polymer blends and composites from renewable resources. Progress in Polymer Science (Oxford). 2006;**31**:576-602

[66] Zhang J, Jiang D, Wilkie C. Polyethylene and polypropylene nanocomposites based upon an oligomerically modified clay. Thermochimica Acta. 2005;**430**:107-113

[67] Deka B, Maji T. Study on the properties of nanocomposite based on high density polyethylene, polypropylene, polyvinyl chloride and wood. Composites Part A: Applied Science and Manufacturing. 2011;**42**: 686-693

[68] Freudenberg K. The relation of cellulose to lignin in wood. Journal of Chemical Education. 1932;**9**:1171-1180

[69] Lee K. On the use of nanocellulose as reinforcement in a polymer matrix. Composites Science and Technology. 2014;**105**:15-27

[70] Saldivar-Guerra E, Vivaldo-Lima E. Handbook of Polymer Synthesis, Characterization, and Processing. 3rd ed. New York: John Wiley & Sons; 2013. pp. 11-12

[71] Robeson L. Polymer Blends: A Comprehensive Review. Munich, Cincinnati: Hanser; 2007

[72] Flory P. Thermodynamics of high polymer solutions. Journal of Chemical Physics. 1941;**9**:440-440

[73] Seuring J, Agarwal S. Polymers with upper critical solution temperature in aqueous solution. Macromolecular Rapid Communications. 2012;**33**:1898-1920

[74] Zhang Q, Hoogenboom R. Polymers with upper critical solution temperature behaviour in alcohol/water solvent mixtures. Progress in Polymer Science. 2015;**48**:122-142

[75] Müller M. Phase behaviour and chain conformations in polymer blends: Monte Carlo simulation vs mean field theory, computational soft matter: From synthetic polymers to proteins. Lecture Notes. 2004;**23**

[76] Fekete E, Foldes E, Damsits F, Pukanszky B. Interaction-structure-property relationships in amorphous polymer blends. Polymer Bulletin. 2000; **44**:363-370

[77] Weeks N, Karasz F, MacKnight W. Enthalpy of mixing of poly(2,6-dimethyl phenylene oxide) and polystyrene. Journal of Applied Physics. 1977;**48**: 4068-4071

[78] Yokouchi M, Seto S, Kobayashi Y. Comparison of polystyrene, poly (styrene/acrylonitrile), high-impact polystyrene, and poly(acrylonitrile/ butadiene/styrene) with respect to tensile and impact properties. Journal of Applied Polymer. 1983;**28**:2209-2216

[79] Sundararaj U, Macosko C. Drop breakup and coalescence in polymer blends: The effects of concentration and Compatibilization. Macromolecules. 1995;**28**:2647-2657

[80] Utracki L. Compatibilization of polymer blends. The Canadian Journal of Chemical Engineering. 2002;**80**: 1008-1016

[81] Chen C, White J. Compatibilizing agents in polymer blends: Interfacial tension. Phase Morphology, and Mechanical Properties, Polymer Science and Engineering. 1993;**33**:923-930

[82] Aziz A, Akil H, Jamaludin S, Ramli N. The effect of multiple compatibilizers on the impact properties of polypropylene/polystyrene (PP/PS) blend. Polymer—Plastics Technology and Engineering. 2011;**50**:768-775

[83] Singh A, Rajiv P, Dhananjai P. A comparative thermal, optical, morphological and mechanical properties studies of pristine and C15A nanoclay-modified PC/PMMA blends: A critical evaluation of the role of nanoclay particles as compatibilizers. The Royal Society of Chemistry. 2013;**3**: 15411-15420

[84] Kasaliwal G, Goldel A, Potschke P. Influence of processing conditions in small-scale melt mixing and compression moulding on the resistivity and morphology of polycarbonate—MWNT composites. Journal of Applied Polymer Science. 2009;**112**:3494-3509

[85] Pandikumar A, Soon How GT, See TP. Graphene and its nanocomposite

material based electrochemical sensor platform for dopamine. RSC Advances. 2014;**4**:63296-63323

[86] Upadhyay R, Soin N, Roy S. Role of graphene/metal oxide composites as photocatalysts, adsorbents and disinfectants in water treatment: A review. RSC Advances. 2014;**4**:3823-3851

[87] Taniike T, Goto K, Terano M. Active site nature of magnesium dichloride-supported titanocene catalysts in olefin polymerization. 2015;**2**:57-63

[88] Deka B, Maji T. Effect of coupling agent and nanoclay on properties of HDPE, LDPE, PP, PVC blend and Phargamites karka nanocomposite. Composites Science and Technology. 2010;**70**:1755-1761

[89] Natta G, Pino P, Mazzanti P. Regular linear head-to-tail polymerizates of certain unsaturated hydrocarbons and filaments comprising said polymerizates. 1973. U.S. Patent. 3715344

[90] Sonnier R, Massardier V, Cassagnau P. Compatibilization of hiPP/HDPE blends by a metallocene copolymer. Journal of Materials. 2008;**1–4**

[91] Shamiri A, Chakrabarti M, Jahan S, Hussain M, Kaminsky W, Aravind P, et al. The influence of Ziegler-Natta and metallocene catalysts on polyolefin structure, properties and processing ability. Materials. 2014;**7**:5069-5108

[92] Wildes G, Keskkula H, Paul D. Coalescence in PC/SAN blends: Effect of reactive compatibilization and matrix phase viscosity. Polymer. 1999;**40**: 5609-5621

[93] Robeson L. Historical perspective of advances in the science and technology of polymer blends. Polymers. 2014;**6**: 1251-1265

[94] Roe R. Use of Block Copolymer as Polymer Blend Compatibilizer. Durham, NC: U.S. Army Research Office; 1993

[95] Rudin A, Phillip C. The Elements of Polymer Science and Engineering. 3rd ed. Oxford: Academic Press; 2013

[96] Tan N. Reactive Compatibilization in Immiscible Polymer Blends. Doctor of Philosophy Thesis: University of Maryland; 1994

[97] Si M, Araki T, Ade H, Kilcoyne L, Sokolov J, Rafailovich M, et al. Compatibilizing bulk polymer blends by using Organoclays compatibilizing bulk polymer blends by using Organoclays. Macromolecules. 2006;**39**:4793-4801

[98] Li B, Wan C, Zhang Y, Ji J. Blends of poly(2,6-dimethyl-1,4-phenylene oxide)/ polyamide 6 toughened by maleated polystyrene-based copolymers: Mechanical properties, morphology, and rheology. Journal of Applied Polymer Science. 2010;**115**:3385-3392

[99] Tong J, Huang H, Wu M. Promoting compatibilization effect of graphene oxide on immiscible PS/PVDF blend via water-as sisted mixing extrusion. Composites Science and Technology. 2017;**149**:286-293

[100] Yang J, Feng C, Dai J, Zhang N, Huang T, Wang Y. Compatibilization of immiscible nylon 6/poly(vinylidene fluoride) blends using graphene oxides. Polymer International. 2013;**62**: 1085-1093

[101] Jaemyung K, Laura J, Cote F, Wa Y, Kenneth R, Shull A, et al. Graphene oxide sheets at interfaces. Journal of the American Chemical Society. 2010;**132**: 8180-8186

[102] Dreyer D, Park S, Christopher B, Ruoff R. The chemistry of graphene

oxide. Chemical Society Reviews. 2009; **39**:228-240

[103] Cao Y, Zhang J, Feng J, Wu P. Compatibilization of immiscible polymer blends using graphene oxide sheets. ACS Nano. 2011;**5**:5920-5927

[104] Jiacheng W, Rasheed A, Thuc V, Fawad I. Graphene nanoplatelets in epoxy system: Dispersion, reaggregation, and mechanical properties of nanocomposites. Journal of Nanomaterials. 2015;**12**

[105] Kar G, Biswas S, Bose S. Tailoring the interface of an immiscible polymer blend by a mutually miscible homopolymer grafted onto graphene oxide: Outstanding mechanical properties. Physical Chemistry Chemical Physics. 2015;**17**:1811-1821

[106] Amani M, Sharif M, Kashkooli A, Rahnama N, Fazli A. Effect of mixing conditions on the selective localization of graphite oxide and the properties of polyethylene/high-impact polystyrene/ graphite oxide nanocomposite blends. RSC Advances. 2015;**5**:77723-77777

[107] Tarfaoui M, Lafdi K, Beloufa I, Daloia D, Muhsan A. Effect of graphene nano-additives on the local mechanical behavior of derived polymer nanocomposites. Polymers. 2018;**10**:667

[108] Niyitanga T, Hae J. Electrochemical properties of poly(vinyl alcohol) and graphene oxide composite for supercapacitor applications. Chemical Physics Letters. 2017;**669**:125-129

[109] Won J, Richard S, Piner D, An J, Ruoff R. American chemical. Society. 2010;**4**:6557-6564

[110] Moon I, Lee J, Ruoff R, Lee H. Reduced graphene oxide by chemical graphitization, nature. Communications. 2010;**1**:1-6

[111] Zhanga H, Zhenga W, Yana Q, Yanga Y. Electrically conductive polyethylene terephthalate/graphene nanocomposites prepared by melt compounding. Polymer. 2010;**51**: 1191-1196

[112] Stankovich S, Dikin D, Dommett G, Kohlhaas K, Zimney E, Stach E, et al. Graphene-based composite materials. Nature. 2006;**442**:282-286

[113] Singh R, Kumar R, Singh D. Graphene oxide: Strategies for synthesis, reduction and frontier applications. RSC Advances. 2016;**6**:64993-65011

[114] Aldosari H. The effect of graphene oxide dispersion on structure-property relationships in graphene-based polymer nanocomposites. Journal of Nano Research. 2020;**65**:97-121

[115] Park S, Ruoff R. Chemical methods for the production of graphenes. Nature. 2009;**4**:217-224

[116] Dreyer D, Bielawski C, Ruoff R. The chemistry of graphene oxide. Chemical Society Reviews. 2010;**39**:228-240

[117] Liu J, Tang J, Gooding J. Strategies for chemical modification of graphene and applications of chemically modified graphene. Journal of Materials Chemistry. 2012;**22**:12435

[118] Liu J, Yang W, Tao L. Thermosensitive graphene nanocomposites formed using pyrene-terminal polymers made by RAFT polymerization. Journal of Polymer Science Part A: Polymer Chemistry. 2010;**48**:425-433

[119] Ibrahim A, Klopocinska A, Horvat K, Abdel Hamid Z. Graphene-based

nanocomposites: Synthesis, mechanical properties, and characterizations. Polymers. 2021;**13**:2869

[120] Duncan T. Applications of nanotechnology in food packaging and food safety: Barrier materials, antimicrobials and sensors. Journal of Colloid and Interface Science. 2011;**363**: 1-24

[121] Cui Y, Kundalwal S, Kumar S. Gas barrier performance of graphene/ polymer nanocomposites. Carbon. 2016; **98**:313-333

Chapter 4

# Chemical Vapor Deposition Synthesis of Graphene on Copper Foils

*Ali Roberto Ruiz Hernández, Adrián Gutierrez Cruz and Jessica Campos-Delgado*

## Abstract

Chemical vapor deposition (CVD) represents a viable synthesis route to produce good-quality, large-area graphene films. In simple words, the technique relies on the thermal decomposition of a carbon-rich source and the further deposition of carbon atoms in a honeycomb pattern on top of a metallic catalyst film. Due to the versatility of the method, many alternatives have been explored for the synthesis of this amazing carbon 2D nanomaterial: low pressure, atmospheric pressure, roll-to-roll. Different catalysts have been explored as well; however, copper (Cu) represents the prime choice, being micrometer-thick foils the most commonly used form in CVD experiments. This chapter focuses on the production of graphene *via* the CVD method using copper foils, and it commences by explaining the generalities of the technique and its variants; next, a description of the method for the production of graphene using copper is included as well as the different precursors (gas, liquid, solid) that have been reported for its synthesis; we continue explaining the importance of the other gases involved in the synthesis and the efforts toward production of large-size single crystals; the obliged transfer process is reviewed, and we conclude by analyzing the advantages and the challenges of the technique.

**Keywords:** CVD, graphene, polycrystalline, copper foil, precursor

## 1. Introduction

Graphene belongs to the extensive realm of nanostructured allotropes of carbon. It is a 2D nanomaterial conformed of $sp^2$-bonded carbon atoms arranged in a hexagonal pattern, each carbon atom bonds in the plane to other three atoms, forming a chicken-wire network only one-atom thick. This amazing nanomaterial was isolated in 2004 by Novoselov and Geim [1], who later, in 2010, were awarded the Nobel Prize in Physics. Perhaps the most natural way to obtain graphene is by exfoliating a layer from graphite, such was the method used by Novoselov and Geim. This method, however, poses intrinsic disadvantages such as the low yield, the time required to find monocrystals, and the irreproducibility of the shapes of the crystals. Scientists realized that a method to produce graphene in a more controlled, reproducible, and scalable way was necessary. Among the many techniques available to produce nanomaterials, the chemical vapor deposition (CVD) method has proven to be advantageous for the production of carbon nanostructures, it has been used to produce carbon nanotubes, carbon nanospheres, nanoribbons, fullerenes, and in 2009, the production of multilayer graphene (MLG) was reported on top of nickel (Ni) films [2, 3]. Later the same year, the synthesis of large-area graphene films was reported using Cu foils by the group of R.S. Ruoff [4], ever since, most CVD-graphene reports use copper foils as catalysts. The basics of the technique, the roles of the precursor, the catalysts, and the synthesis gases are reviewed in detail in the following paragraphs.

Early reports of CVD-graphene accounted for continuous films with regions of single-layer, bilayer, and few-layer graphene (SLG, BLG, FLG, respectively). Characterization of these films revealed that the BLG and FLG do not respect an AB stacking of the layers, the stacking from highly oriented pyrolytic graphite (HOPG); the orientation between adjacent layers is random in CVD-grown films. Another peculiarity of these graphene films is their polycrystallinity, as will be described below in detail.

## 2. Chemical vapor deposition

The chemical vapor deposition technique is based on the chemical thermal decomposition of a precursor (the compound that provides the feedstock for the production of the desired material), so their individual constituents can build up a solid film or nanostructure into a specified substrate. In general, a common CVD equipment is composed of three main modules: 1) a precursor injection module, 2) the reaction site or chamber, and 3) the gas ejection module [5]. Although these three modules can be arranged either vertically or horizontally, the horizontal configuration is the most widely used. The reaction site consists of a region subjected to high temperatures where the thermal decomposition of the precursor can be triggered. A very common setup consists of a tubular furnace with a borosilicate/quartz/alumina tube inside serving as reaction chamber, see **Figure 1**, the nature of the tube would depend on the temperature conditions used in the process. If the process requires a catalyst, it can be placed inside the reaction chamber, such is the case for the CVD synthesis of graphene as will be reviewed in detail below. Depending on the mechanism used for the delivery and the type of precursor, the pressure conditions employed, different CVD configurations can be considered [5]. Each one of these alternatives can undergo the same tasks, but each will use a different approach to fulfill it. In the following sections, some of the most common and popular configurations will be exposed.

### 2.1 APCVD

The atmospheric pressure chemical vapor deposition (APCVD) is one of the simplest configurations found for a CVD system since it operates in normal conditions (atmospheric pressure), making this option more affordable and easier to implement at laboratories. However, due to these conditions, problems related to the dispersion of the gases within the chamber have been considered as one of its major disadvantages, because the decomposed particles will not cover the substrate homogeneously as expected [6, 7].

For this approach, different precursors can be used. In the case of liquids or solids dissolved in solution, a bubbler or an evaporation system will be required to stimulate the particles to be expulsed and consequently be transferred into the chamber by the action of a carrier gas, provoking its decomposition in the reaction site [6].

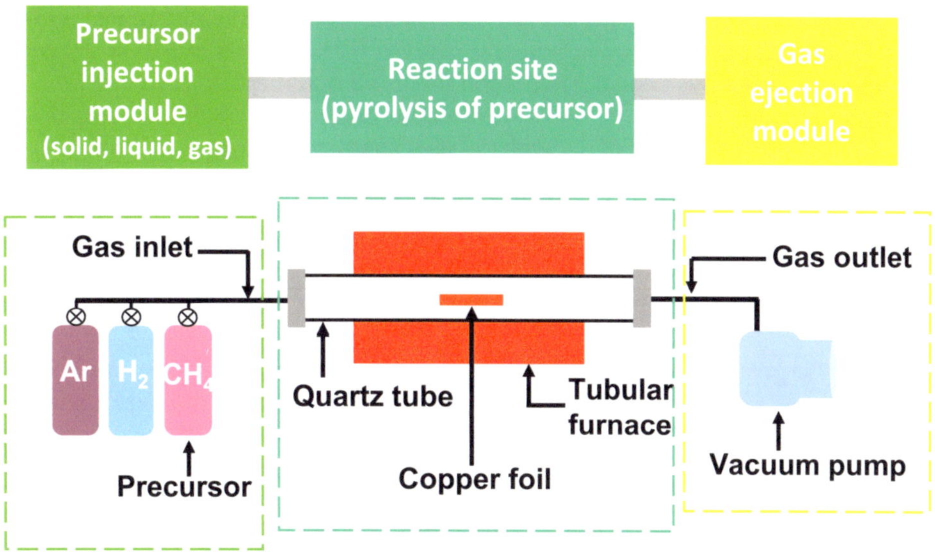


**Figure 1.**
*Schematic representation of a horizontal CVD setup, specifying the three constituent modules (top) and diagram of an LPCVD setup for the synthesis of graphene on copper foil and methane as precursor (bottom), dashed rectangles highlight the corresponding modules.*

### 2.2 LPCVD

The low-pressure chemical vapor deposition (LPCVD) is based on the use of vacuum for the outgassing of the system to achieve low pressures throughout the deposition process. In comparison with the last approach, with this technique, the dispersion and the deposition are enhanced, making it possible to obtain better coverage within the substrates and a higher quality. The pressures commonly used are between $10^{-1}$ and $10^{-2}$ torr and are reached with the use of mechanical and molecular vacuum pumps connected to the gas ejection module [6].

### 2.3 AACVD

The aerosol-assisted chemical vapor deposition (AACVD) is characterized for using fine sub-micrometer-sized aerosol droplets of precursor dispersed throughout a gaseous medium, which eventually are transported into the reaction zone to be decomposed and deposited in the target substrate. The atomization of the liquid precursor can be achieved by different routes, as it could be the use of an ultrasonic aerosol generation, a pneumatic aerosol jet, or *via* an electrostatic atomization. The AACVD possesses the advantage that thermally unstable and nonvolatile precursors can be employed for the synthesis of CVD products at lower cost. In addition, this variant is considered a more flexible option since it can be carried out under low pressure, atmospheric pressure, and in open atmosphere [8].

### 2.4 Roll-to-roll CVD

Roll-to-roll CVD is one of the most recent approaches developed, and it consists of a continuous process that can carry out the deposition of the material on a substrate and its withdrawal or a sequential process such as a further transference into another suitable substrate (in some cases it is required). This CVD variant is equipped with conveyor belts that are in charge of the transport of the substrates into the reaction site, where a continuous inlet of gases throughout the chamber leads to a constant deposit built up in the incoming substrate without interruption. Subsequently, the conveyor belts either introduce the material into a second stage, where a transfer procedure can be done automatically, or take the substrate out of the system. This CVD configuration can be developed for different process conditions, such as ambient pressure, low pressure, or a plasma-enhanced CVD [9, 10].

### 2.5 Other configurations of CVD (PECVD, UHV-CVD, CW-CVD)

One of the great advantages of using a CVD system is the wide variety of possible configurations. Above, we have described the most popular ones, but there is still a large catalogue of variants, among which are worth mentioning the plasma-enhanced chemical vapor deposition (PECVD), the ultrahigh-vacuum chemical vapor deposition (UHV-CVD), and the cold-wall chemical vapor deposition (CW-CVD), these alternatives will be briefly discussed in the following lines.

The PECVD variant uses plasma to improve the decomposition of the precursor and its deposition, allowing a reduction in the temperature and the possibility of avoiding the use of a catalytic substrate for the deposit [11]. The UHV-CVD is characterized by the use of more extreme low-pressure conditions ($\sim 10^{-7}$ torr) to avoid the oxidation of the deposit and to enhance the decomposition of the precursor [12]. Finally, the CW-CVD owes its name to the lack of a heating furnace, and it relies on a Joule heating approach for the decomposition of the precursor, achieving a more selective and controllable reaction site, by locally heating the substrate [13, 14].

## 3. Synthesis of graphene *via* CVD

Chemical vapor deposition has proven to be a very efficient method for graphene film synthesis [15]. It is possible to obtain films with few defects, good uniformity, and good control in the number of layers, acquiring films between one and few layers [16, 17].

Regardless of the CVD equipment configuration, almost all graphene synthesis procedures share the following requirements: a transition metal sheet that acts as a catalyst for carbon dissociation and as a substrate for graphene growth, which is placed inside the reaction chamber; a carbon-based precursor material [18]; temperatures around 1,000°C capable of decomposing the precursor material and dissociating the carbon [19, 20], at such temperatures a borosilicate tube is not convenient, the regular choice is a quartz tube, its transparency allows direct monitoring of the interior of the reaction chamber; a flow of hydrogen ($H_2$) that provides a reducing atmosphere and influences growth behavior; and a flow of inert or carrier gas, argon (Ar) being the privileged option [18]. A diagram of a typical LPCVD setup for the synthesis of graphene is depicted in **Figure 1**. A characteristic of CVD-produced graphene films is the polycrystalline nature of the atom-thick membrane, as will be detailed below.

### 3.1 Copper foils as substrates

Although the deposition of graphene on metal substrates by CVD was reported for the first time on other metals such as Ni and platinum (Pt), Cu is currently the most widely used metal as a catalyst substrate for the growth of graphene by CVD [16].

This preference can be explained by the growth mechanism of graphene on these substrates. The difference in mechanisms was elucidated by the group of R.S. Ruoff in 2009 [21] using carbon isotope labeling and Raman spectroscopy mapping of graphene films grown on Ni and Cu. The difference is due to the carbon solubility of each metal. For metals with high solubility (9,000 ppm at 900°C in Ni, and 11,000 ppm at 1,000°C in Pt) [22], carbon atoms infiltrate the metal matrix at high temperatures, and this solubility tends to increase with temperature. When the temperature drops, the solute is precipitated from the matrix, leading to the formation of graphene. This saturation and precipitation process results in the formation of MLG regions [17, 22]. In contrast, for metals with low carbon solubility (7.4 ppm at 1,020°C in Cu) [22], the solubility is relatively low despite high temperatures. Therefore, the process of carbon saturation and precipitation does not occur. Graphene is formed on the surface of Cu, limited to it; hence, the size of the graphene films will be correlated with the copper films in the reaction chamber. This process promotes the formation of SLG [17, 22]. Common Cu foil thicknesses for the growth of graphene films are 25 and 50 micrometers; moreover, the purity of the foil varies among commercial suppliers.

#### *3.1.1 Growth mechanism*

In general, the growth mechanism of graphene by CVD on a copper substrate can be divided into three stages: (1) The precursor molecules collide with the surface of the Cu substrate. As a result, they can be absorbed on the surface, disperse back to the gas phase, or go to the next reaction step. (2) The molecules dehydrogenate totally or partially, or eliminate any element other than carbon (denitrogenation, deoxygenation), and active carbon species are formed. (3) The active species diffuse over the surface, cluster, generate nucleation sites, and begin to grow on graphene islands on the Cu substrate [15, 23]. These graphene islands are grains of honeycomb arranged carbons, they possess a particular orientation with respect to each other, when the grains are large enough and meet carbon atoms from a nearby grain, they coalesce, forming a grain boundary where the mismatch orientation is overcome by the presence of 5-, 6-, 7-, and 8-membered rings [24].

#### *3.1.2 Annealing*

An important step prior to the CVD synthesis of graphene is the thermal treatment of the micrometer-thick Cu foils. Cu substrates undergo an annealing process that consists of heating them at high temperatures close to the Cu melting point, but lower than it (<1,085°C) for a certain period of time [25]. The surface of copper has some roughness, grain boundaries, surface defects, and impurity particles. This process is carried out in order to smooth the copper surface, reduce surface oxides, remove volatile impurities and surface contaminants, and favor the formation of graphene on the surface [26]. In addition, a reorganization of the copper atoms on the surface is promoted, which produces a release of internal stresses and an increase in the size of the Cu microcrystalline structure [25].

Regarding graphene nucleation, flat Cu regions, or terraces, favor the growth of SLG or BLG domains. In contrast, slope regions on the Cu surface, or ledges, promote MLG growth. Additionally, grain boundaries, impurities, and surface defects contribute to the formation of MLG domains [26].

## 3.2 Precursors

The synthesis of graphene using the CVD technique can be carried out with a large catalogue of carbon-based compounds used as precursors. These precursors can be categorized by their physical state or by their chemical structure. By physical state, they can be classified as gaseous, liquid, and solid. While based on its chemical structure, they are separated as aliphatic and aromatic compounds. The use of an appropriate precursor with the right conditions can improve the efficiency of the production process and the quality of the final product [15].

### *3.2.1 Gases*

Gaseous carbon precursors are the main source used for graphene synthesis by CVD. The most used are hydrocarbons such as methane ($CH_4$) and acetylene ($C_2H_2$), followed by ethylene ($C_2H_4$) [27–29]. A gaseous precursor occupies less space than a liquid or solid one, because they can be stored in certain specialized containers. Some precursors are produced as a by-product of industrial processes, such as biogas, which is constituted essentially of methane and carbon dioxide [15].

### *3.2.2 Liquids*

Liquid carbon precursors are less commonly used in CVD methods. The most common precursors used in CVD are hexane ($C_6H_{14}$), ethanol ($C_2H_5OH$), and benzene ($C_6H_6$) [30–32]. These liquid carbon precursors are easy to use and relatively inexpensive compared with gaseous ones [15, 32].

In contrast to the use of gaseous materials, when the precursor used is in a liquid state, a previous stage is needed to transform the precursor from its liquid form to a vapor in order to be deposited on the metal substrate. Various CVD configurations have been used to work with liquid precursors. An approach constitutes the use of an external source of heat to vaporize liquids with high boiling points. Ultrasonic baths can be used to form aerosols from the liquids [15]. Unlike gaseous precursors that have a certain flow, an external constant gas flow is needed to carry the post-treatment precursor to the reaction chamber [15, 16]. Some of these precursors could be harmful to human health, since some may be volatile organic compounds or carcinogenic in nature [15].

### *3.2.3 Solids*

Solid carbon precursors are equally rarely used. They are more complex in terms of chemical and biological structure. Any organic material that works as a carbon source can be used, including plant waste, plastics, animal waste, insect parts, even food. Therefore, solid precursors occupy more space in the reaction chamber than gaseous or liquid precursors [15, 33].

As with liquid precursors, prior to the synthesis process, an extra step is required to convert the solid material into a gas phase before acting on the metal catalyst. More energy is required to carry out the process, which can increase costs. Solid carbon precursors can be placed directly on top of the catalyst metal so that graphene forms on the back of the catalyst. Polymers have been spin coated directly on copper foils and used as carbon precursor [34]. Most of the precursors used are solid wastes or biomass; using these products to form graphene would positively impact waste recycling for the production of a high-value product and may reduce the overall cost of synthesis [15].

**Table 1** presents different carbon precursors that have been used for the synthesis of graphene, divided by their previously discussed physical states, along with a detailed example of the synthesis process using one of those precursors.

| Physical state | Compounds | Example |
|---|---|---|
| Gas | Methane [27], acetylene [28], ethylene [29], propene [35], biogas [36] | Acetylene was used as a precursor gas in a LPCVD to fabricate SLG and BLG at 1,035°C with a flow of $H_2$ [28]. |
| Liquid | Ethanol [32], hexane [30], benzene [31], methanol [37], pyridine [38], 1-propanol [37], 2-phenylethanol [39], palm oil [40] | SLG, BLG and tri-layer graphene (TLG) was synthesized with APCVD using two alcohols as precursors: 2-phenylethanol and ethanol at 980 or 990°C using a Ar-$H_2$ mixture [32]. |
| Solid | PMMA [34], hexabenzocoronene [39], highly oriented pyrolytic graphite [41], cookie [33], cockroach leg [33], grass [33], camphor [42] | APCVD was used to produce SLG and BLG graphene sheets with solid camphor that was evaporated in a first furnace at 200°C, and pyrolyzed in a second furnace at 1,020°C using $H_2$ as carrier gas [42]. |

**Table 1.**
*Carbon precursors used for the synthesis of graphene organized by their physical state. An example of the synthesis performed is shown in each case.*

### *3.2.4 Aliphatic compounds*

To decompose aliphatic precursors, high temperatures are needed to promote the partial or total dehydrogenation of carbon compounds, as well as the removal of other elements that could be previously linked to the chains, such as nitrogen and oxygen. Subsequently, it is necessary to promote the breaking of the carbon-carbon bonds between the chains. This favors the deposition of active carbon species on the metal substrate, the subsequent formation of graphene growth nuclei in $sp^2$ formation, and the capture of active carbon for the growth of the graphene islands [15].

### *3.2.5 Aromatic compounds*

Aromatic compounds require a lower growth temperature, and graphene growth using benzene at around 300°C has been reported by Jang and coworkers [43]. Processes of dehydrogenation and possible denitrogenation and deoxygenation of the aromatic species take place at lower temperatures. Active aromatic ring species are deposited on the substrate forming graphene growth nuclei and capture the active carbon species for island growth. The low energy needed is related to the basic structure of aromatic compounds, which is analogous to the hexagonal honeycomb packaging of graphene [15].

### *3.2.6 Greener options*

In order to develop ecological and environmentally friendly CVD methods, green and bio-renewable carbon sources that are easily obtained at low cost are sought [40, 42, 44].

With the implementation of these precursors, it is attempted to reduce the use of greenhouse gases, the exploitation of toxic materials, and the development of ecological and sustainable technologies [44].

It has been possible to synthesize graphene using various ecological precursors such as foods, for example, cookies, chocolate, honey, sugar, butter, milk, cheese; waste such as plastics, grass, bones, eggshells, dog and cow feces, wood, leaf, bagasse, fruit, tea tree extracts; and derivatives of insects [44].

In particular, the use of solid botanical camphor ($C_{10}H_{16}O$) for the synthesis of SLG and BLG graphene sheets has been studied. Camphor is a natural solid botanical hydrocarbon source, which is regenerative, low-cost, and environmentally friendly [42]. Similarly, the use of palm oil as a carbon precursor for the synthesis of MLG has been studied. Palm oil is a natural oil source that has a unique chemical composition with long-chain carbon. It has high potential as a green and renewable carbon precursor in the large-scale production of graphene [40].

### 3.3 Gases

Molecular hydrogen has a fundamental role in the synthesis of graphene by CVD since it acts as an activating gas. It acts by cleaning the surface of the substrate, reducing impurities and defects during the annealing process, as well as reducing surface oxides [17, 23]. It behaves as a co-catalyst, along with the substrate, promoting the growth of graphene, as well as an etchant to control the growth and its properties [23, 45]. It controls the adsorption, stability, thickness or number of layers, the population on the catalyst surface, the configuration of the edges, and the morphology of the domains of the grown graphene [17, 23, 45].

The flow of $H_2$ has been considered as an important factor for the quality of graphene during synthesis. By using a gaseous precursor, the radius of $H_2$ with respect to the precursor gas can be studied [17]. It has been observed that when the $H_2/CH_4$ volume ratio in a process is greater than 0.5, the graphene remains in small separate islands and does not coalesce to form a large film. This may happen because excess $H_2$ can limit graphene growth by removing weak carbon-carbon bonds despite acting as a co-catalyst [46]. If the ratio of $H_2$ to $CH_4$ is too high, the etching of carbon species becomes much faster than the formation of graphene layers [47], destroying the integrity of the network and reducing the quality of graphene [17].

Regarding the flow of the precursor gas, it has been reported that, for methane, the number of layers does not depend on the $CH_4$ flow rate, although an increase in defects was observed with the increase in flow [17].

Similarly, the use of an inert gas can help the synthesis process. The inert gas can be used to dilute the carbon feedstock to achieve high $H_2$ to precursor ratios, to increase the total pressure of the reaction chamber, or to dilute flammable or explosive material below its lower explosive limit [18, 46]. The concentration of each gas can range between 0 and 100%. The explosive limits of $H_2$ in air range between 18 and 60%, and the flammable limits between 4 and 75%. Pure $H_2$ and $H_2$ mixture beyond the flammable limit are dangerous [46]. Furthermore, it can be used as a carrier gas for precursor molecules in certain cases [16]. Gases such as Ar and nitrogen ($N_2$) have been used to fulfill this function [16, 18, 46]. Ar is chemically inert under the conditions used during the synthesis [18, 46]. $N_2$ is inert as well, although more abundant and cheaper than Ar. Since it exists as a diatomic molecule, there is a risk that $N_2$ will dissociate during synthesis. The dissociation product could act as a substituent in the graphene film, causing a doping or interruption of the network. However, the triple bond present has a very large binding energy, so the rate of $N_2$ dissociation should be extremely small at the temperatures used. Furthermore, Cu is an inefficient catalyst for the dissociation of $N_2$ [18].

## 4. Large-area single-crystal graphene

Using a CVD system, large-area SLG can be synthesized with high yields; however, its crystallinity can represent an important disadvantage for its application. The polycrystalline nature of the Cu substrate and the presence of impurities within its surface lead to a high density of nucleation sites and adlayers, resulting in a polycrystalline material that will present diminished mechanical and electrical properties than the ideal product.

For this reason, in recent years, different alternatives have been proposed to solve this important factor. Li *et al.* and Yang *et al.* presented a methodology to reduce the nucleation sites during the growth, based on the use of Cu foil enclosures (folded as little pockets) that have the purpose to reduce the partial pressure of the precursor and undesired species inside the cavity of the enclosure, such as $SiO_2$ particles proceeding from the decomposition of the quartz reaction tube, commonly found as contaminants in the graphene surface [48, 49], achieving single crystals in the size range of several hundreds of microns. On the other hand, other research groups have used other alternatives, such as Weatherup *et al.* and Huang *et al.* employed metal catalyst alloys (e.g., Au, Cu/Ni(III)) to grow high-quality single-crystal graphene, also by reducing the nucleation density [50, 51]. In addition, the action of oxygen ($O_2$) as a reducer of nucleation has been also studied. Hao and coworkers found in 2013 that $O_2$ can act as a passivation agent for the nucleation sites, reducing in consequence the density of the adlayers produced. They also concluded that $O_2$ can be provided by an external source or even by segregation within the Cu substrate, such as a Cu oxide [52]. Now, millimeter-sized single crystals have been achieved [53], this holds promise for the scalable and affordable production of graphene for electronic devices.

## 5. Transfer procedures

Once graphene has been synthesized on Cu foils, its transfer to more suitable substrates for its study or to be used in different applications is necessary. There are several procedures that have been developed for this transfer step, having as purposes: a) to separate the graphene film from the original metal substrate where it was deposited and b) to protect the graphene after the transference [54]. In general, these procedures can be classified in two groups, the wet chemical methods and the dry chemical methods.

### 5.1 Wet chemical methods

These methods are based generally on the coating of a protective polymeric layer over the graphene film (e.g., PDMS or PMMA), followed by the immerse in an etching solvent (e.g., $FeCl_3$, HCl, $HNO_3$, $Fe(NO_3)_3$, $CuCl_2$) to remove the metal substrate, so the graphene can be positioned then in another substrate (e.g., $SiO_2$) [55]. **Figure 2** depicts a flowchart of the transfer process using PMMA. Despite the good results that could be obtained with this general technique, in recent years, new variants have been developed to overcome its disadvantages (i.e., the presence of impurities and the formation

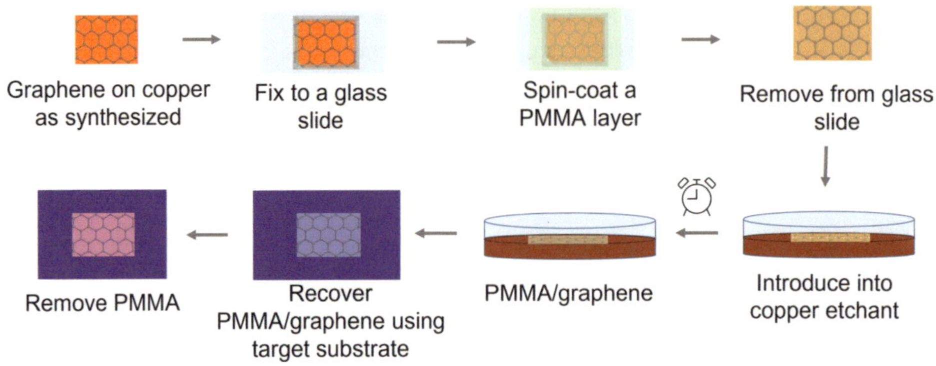


**Figure 2.**
*Schematic representation of the wet transfer process using spin-coated PMMA.*

of cracks in the material). Such is the case of An and collaborators, who developed a reverse transfer technique by locating the graphene/PMMA bilayer into a flexible substrate (e.g., PET), reducing the density of cracks and the presence of impurities [56]. Lin and coworkers found a way to avoid the use of the polymeric layer, achieving better quality of transferred graphene (free of organic residues). They used isopropyl alcohol (IPA) and ammonium persulfate solution ($(NH_4)_2S_2O_8$) as the etchant agent, which was later substituted with DI water and IPA to reduce the surface tension and ease the transference to the desired substrate [57]. Furthermore, the need of the etching step has been also avoided in some techniques, such as Chandrashekar *et al.* reported a method to separate the Cu foil by oxidizing it in hot water, modifying its nature from hydrophobic to hydrophilic, easing its delamination from graphene [58].

### 5.2 Dry chemical methods

The dry chemical methods are based on the same principle as the wet ones, but with the difference that these ones are performed under dry conditions. Several methodologies have been proposed as well, such as Yang *et al.* reported an electrochemical procedure for the etching of the Cu substrate to reduce the contamination in graphene [59]. Martins′ group developed a transference methodology *via* lamination into flexible substrates, using a direct transfer procedure without intermediates [60].

## 6. Conclusions

The chemical vapor deposition method provides graphene films of good quality and large area, the many variants of the method allow versatility in a number of parameters such as: precursor used, temperature, and pressure. From its first reports more than 10 years ago, CVD has remained as reliable technique to produce graphene films, many advances have been reported since, and understanding of the role of every parameter in the synthesis has been achieved. CVD graphene is polycrystalline by nature, and when more than one layer is produced, the mismatch or stacking of layers is not controlled, and these characteristics might condition its performance at the mechanical and electrical level. Applications that require continuous, conductive, transparent films can be tackled using CVD graphene; however, the transfer process remains a challenge to overcome.

## Author details

Ali Roberto Ruiz Hernández, Adrián Gutierrez Cruz and Jessica Campos-Delgado*
University of the Americas Puebla, San Andrés Cholula, Puebla, Mexico

*Address all correspondence to: jessica.mex@gmail.com

## References

[1] Novoselov KS, Geim AK, Morozov SV, Jiang D, Zhang Y, Dubonos SV, et al. Electric field effect in atomically thin carbon films. Science. 2004;**306**:666-669. DOI: 10.1126/science.1102896

[2] Reina A, Jia X, Ho J, Nezich D, Son H, Bulovic V, et al. Large area, few-layer graphene films on arbitrary substrates by chemical vapor deposition. Nano Letters. 2009;**9**:30-35. DOI: 10.1021/nl801827v

[3] Kim KS, Zhao Y, Jang H, Lee SY, Kim JM, Kim KS, et al. Large-scale pattern growth of graphene films for stretchable transparent electrodes. Nature. 2009;**457**:706-710. DOI: 10.1038/nature07719

[4] Li X, Cai W, An J, Kim S, Nah J, Yang D, et al. Large-area synthesis of high-quality and uniform graphene films on copper foils. Science. 2009;**324**:1312-1314. DOI: 10.1126/science.1171245

[5] Carlsson J, Martin PM. Chemical vapor deposition. Handbook of Deposition Technologies for Films and Coatings. 2010;**1**:314-363. DOI: 10.1016/b978-0-8155-2031-3.00007-7

[6] Pottathara YB, Grohens Y, Kokol V, Kalarikkal N, Thomas S. Synthesis and processing of emerging two-dimensional nanomaterials. In: Pottathara Y, Thomas S, Kalarikkal N, Grohens Y, Kokol V, editors. Nanomaterials Synthesis. New York: Elsevier; 2019. pp. 1-25. DOI: 10.1016/B978-0-12-815751-0.00001-8

[7] Vlassiouk I, Fulvio P, Meyer H, Lavrik N, Dai S, Datskos P, et al. Large scale atmospheric pressure chemical vapor deposition of graphene. Carbon. 2013;**54**:58-67. DOI: 10.1016/j.carbon.2012.11.003

[8] Hou X, Choy KL. Processing and applications of aerosol-assisted chemical vapor deposition. Chemical Vapor Deposition. 2006;**12**:583-596. DOI: 10.1002/cvde.200600033

[9] Bae S, Kim H, Lee Y, Xu X, Park JS, Zheng Y, et al. Roll-to-roll production of 30-inch graphene films for transparent electrodes. Nature Nanotechnology. 2010;**5**:574-578. DOI: 10.1038/nnano.2010.132

[10] Xin H, Li W. A review on high throughput roll-to-roll manufacturing of chemical vapor deposition graphene. Applied Physics Reviews. 2018;**5**:031105. DOI: 10.1063/1.5035295

[11] Yi K, Liu D, Chen X, Yang J, Wei D, Liu Y, et al. Plasma-enhanced chemical vapor deposition of two-dimensional materials for applications. Accounts of Chemical Research. 2021;**54**:1011-1022. DOI: 10.1021/acs.accounts.0c00757

[12] Meyerson BS. Low-temperature silicon epitaxy by ultrahigh vacuum/chemical vapor deposition. Applied Physics Letters. 1986;**48**:797-799. DOI: 10.1063/1.96673

[13] Bointon TH, Barnes MD, Russo S, Craciun MF. High quality monolayer graphene synthesized by resistive heating cold wall chemical vapor deposition. Advanced Materials. 2015;**27**:4200-4206. DOI: 10.1002/adma.201501600

[14] Alnuaimi A, Almansouri I, Saadat I, Nayfeh A. Toward fast growth of large area high quality graphene using a cold-wall CVD reactor. RSC Advances. 2017;**7**:51951-51957. DOI: 10.1039/C7RA10336K

[15] Kairi MI, Khavarian M, Bakar SA, Vigolo B, Mohamed AR. Recent trends in

graphene materials synthesized by CVD with various carbon precursors. Journal of Materials Science. 2018;**53**:851-879. DOI: 10.1007/s10853-017-1694-1

[16] Saeed M, Alshammari Y, Majeed SA, Al-Nasrallah E. Chemical vapour deposition of graphene-synthesis, characterisation, and applications: A review. Molecules. 2020;**25**:3856. DOI: 10.3390/molecules25173856

[17] Wang C, Vinodgopal K, Dai G. Large-area synthesis and growth mechanism of graphene by chemical vapor deposition. In: Mandracci P, editor. Chemical Vapor Deposition for Nanotechnology. 1st ed. London: IntechOpen; 2018. pp. 97-113. DOI: 10.5772/intechopen.79959

[18] Gausden J, Siris R, Stimpel-Lindner T, McEvoy N, Duesberg GS, Hallam T. Nitrogen as a suitable replacement for argon within methane-based hot-wall graphene chemical vapor deposition. Physica Status Solidi B. 2019;**256**:1900240. DOI: 10.1002/pssb.201900240

[19] Liu Z, Lin L, Ren H, Sun X. CVD synthesis of graphene. In: Zhang G, editor. Thermal Transport in Carbon-Based Nanomaterials. 1st ed. Amsterdam: Elsevier; 2017. pp. 19-56. DOI: 10.1016/B978-0-32-346240-2.00002-9

[20] Feng Y, Trainer DJ, Peng H, Liu Y, Chen K. Safe growth of graphene from non-flammable gas mixtures via chemical vapor deposition. Journal of Materials Science and Technology. 2017;**33**:285-290. DOI: 10.1016/j.jmst.2016.09.024

[21] Li X, Cai W, Colombo L, Ruoff RS. Evolution of graphene growth on Ni and Cu by carbon isotope labeling. Nano Letters. 2009;**9**:4268-4272. DOI: 10.1021/nl902515k

[22] Xu S, Zhang L, Wang B, Ruoff RS. Chemical vapor deposition of graphene on thin-metal films. Cell Reports Physical Science. 2021;**2**:100372. DOI: 10.1016/j.xcrp.2021.100372

[23] Jin Y, Hu B, Wei Z, Luo Z, Wei D, Xi Y, et al. Roles of H2 in annealing and growth times of graphene CVD synthesis over copper foil. Journal of Materials Chemistry A. 2014;**2**:16208-16216. DOI: 10.1039/C4TA02557A

[24] Huang PY, Ruiz-Vargas CS, van der Zande AM, Whitney WS, Levendorf MP, Kevek JW, et al. Grains and grain boundaries in single-layer graphene atomic patchwork quilts. Nature. 2011;**469**:389-392. DOI: 10.1038/nature09718

[25] Kondrashov I, Komlenok M, Pivovarov P, Savin S, Obraztsova E, Rybin M. Influence of different copper treatment on the formation of single-layer graphene by CVD method. In: The 2nd International Online-Conference on Nanomaterials. 2021

[26] Ibrahim A, Akhtar S, Atieh M, Karnik R, Laoui T. Effects of annealing on surface morphology of copper substrate and graphene growth by chemical vapor deposition. Carbon. 2015;**94**:369-377. DOI: 10.1016/j.carbon.2015.06.067

[27] Kordatos A, Kelaidis N, Giamini SA, Velasco JM. AB stacked few layer graphene growth by chemical vapor deposition on single crystal Rh(111) and electronic structure characterization. Applied Surface Science. 2016;**369**:251-256. DOI: 10.1016/j.apsusc.2016.02.023

[28] Yang M, Sasaki S, Suzuki K, Miura H. Control of the nucleation and quality of graphene grown by low- pressure chemical vapor deposition with acetylene. Applied Surface Science.

2016;**366**:219-226. DOI: 10.1016/j.apsusc.2016.01.089

[29] Trinsoutrot P, Vergnes H, Caussat B. Three dimensional graphene synthesis on nickel foam by chemical vapor deposition from ethylene. Materials Science and Engineering B. 2014;**179**:12-16. DOI: 10.1016/j.mseb.2013.09.018

[30] Srivastava A, Galande C, Ci L, Song L, Rai C, Jariwala D, et al. Novel liquid precursor-based facile synthesis of large-area continuous, single, and few-layer graphene films. Chemistry of Materials. 2010;**22**:3457-3461. DOI: 10.1021/cm101027c

[31] Li Z, Wu P, Wang C, Fan X, Zhang W, Zhai X, et al. Low-temperature growth of graphene by chemical vapor deposition using solid and liquid carbon sources. ACS Nano. 2011;**5**:3385-3390. DOI: 10.1021/nn200854p

[32] Campos-Delgado J, Botello-Méndez AR, Algara-Siller G, Hackens B, Pardoen T, Kaiser U, et al. CVD synthesis of mono- and few-layer graphene using alcohols at low hydrogen concentration and atmospheric pressure. Chemical Physics Letters. 2013;**584**:142-146. DOI: 10.1016/j.cplett.2013.08.031

[33] Ruan G, Sun Z, Peng Z, Tour JM. Growth of graphene from food, insects and waste. ACS Nano. 2011;**5**:7601-7607. DOI: 10.1021/nn202625c

[34] Hussain A, Mehdi SM, Abbas N, Hussain M, Naqvi RA. Synthesis of graphene from solid carbon sources: A focused review. Materials Chemistry and Physics. 2020;**248**:122924. DOI: 10.1016/j.matchemphys.2020.122924

[35] Gotterbarm K, Zhao W, Höfert O, Gleichweit C, Papp C, Steinrück HP. Growth and oxidation of graphene on Rh(111). Physical Chemistry Chemical Physics. 2013;**15**:19625-19631. DOI: 10.1039/C3CP53802H

[36] Strudwick AJ, Weber NE, Schwab MG, Kettner M, Weitz RT, Wünsch JR, et al. Chemical vapor deposition of high quality graphene films from carbon dioxide atmospheres. ACS Nano. 2015;**9**:31-42. DOI: 10.1021/nn504822m

[37] Guermoune A, Chari T, Popescu F, Sabri SS, Guillemette J, Skulason HS, et al. Chemical vapor deposition synthesis of graphene on copper with methanol, ethanol, and propanol precursors. Carbon. 2011;**49**:4204-4210. DOI: 10.1016/j.carbon.2011.05.054

[38] Xue Y, Wu B, Jiang L, Guo Y, Huang L, Chen J, et al. Low temperature growth of highly nitrogen-doped single crystal graphene arrays by chemical vapor deposition. Journal of the American Chemical Society. 2012;**134**:11060-11063. DOI: 10.1021/ja302483t

[39] Gan X, Zhou H, Zhu B, Yu XY. A simple method to synthesize graphene at 633 K by dechlorination of hexachlorobenzene on Cu foils. Carbon. 2012;**50**:306-310. DOI: 10.1016/j.carbon.2011.08.057

[40] Salifairus MJ, Abd Hamid SB, Soga T, Alrokayan SAH, Khan HA, Rusop M. Structural and optical properties of graphene from green carbon source via thermal chemical vapor deposition. Journal of Materials Research. 2016;**31**:1947-1956. DOI: 10.1557/jmr.2016.200

[41] Xu M, Fujita D, Sagisaka K, Watanabe E, Hanagata N. Production of extended single-layer. ACS Nano. 2011;**5**:1522-1528. DOI: 10.1021/nn103428k

[42] Kalita G, Wakita K, Umeno M. Monolayer graphene from a green solid precursor. Physica E: Low-dimensional Systems and Nanostructures. 2011;**43**:1490-1493. DOI: 10.1016/j.physe.2011.04.014

[43] Jang J, Son M, Chung S, Kim K, Cho C, Lee BH, et al. Low-temperature grown continuous graphene films from benzene by chemical vapor deposition at ambient pressure. Scientific Reports. 2015;**5**:17955. DOI: 10.1038/srep17955

[44] Kumar R, Kumar Singh R, Pratap SD. Natural and waste hydrocarbon precursors for the synthesis of carbon based nanomaterials: Graphene and CNTs. Renewable and Sustainable Energy Reviews. 2016;**58**:976-1006. DOI: 10.1016/j.rser.2015.12.120

[45] Li X, Li X, Zang X, Zhu M, He Y, Wang K, et al. Role of hydrogen in the chemical vapor deposition growth of $MoS_2$ atomic layers. Nanoscale. 2015;**7**:8398-8404. DOI: 10.1039/C5NR00904A

[46] Feng Y, Trainer DJ, Peng H, Liu Y, Chen K. Safe growth of graphene from non-flammable gas mixtures via chemical vapor deposition. Journal of Materials Science and Technology. 2017;**33**:285-290. DOI: 10.1016/j.jmst.2016.09.024

[47] Park HJ, Meyer J, Roth S, Skákalová V. Growth and properties of few-layer graphene prepared by chemical vapor deposition. Carbon. 2010;**48**:1088-1094. DOI: 10.1016/j.carbon.2009.11.030

[48] Li X, Magnuson CW, Venugopal A, Tromp RM, Hannon JB, Vogel EM, et al. Large-area graphene single crystals grown by low-pressure chemical vapor deposition of methane on copper. Journal of the American Chemical Society. 2011;**133**:2816-2819. DOI: 10.1021/ja109793s

[49] Yang M, Sasaki S, Ohnishi M, Suzuki K, Miura H. Electronic properties and strain sensitivity of CVD-grown graphene with acetylene. Japanese Journal of Applied Physics. 2016;**55**:04EP05

[50] Weatherup RS, Bayer BC, Blume R, Ducati C, Baehtz C, Schloogl R, et al. In situ characterization of alloy catalysts for low-temperature graphene growth. Nano Letters. 2011;**11**:4154-4160. DOI: 10.1021/nl202036y

[51] Huang M, Biswal M, Park HJ, Jin S, Qu D, Hong S, et al. Highly oriented monolayer graphene grown on a Cu/Ni (111) alloy foil. ACS Nano. 2018;**12**:6117-6127. DOI: 10.1021/acsnano.8b02444

[52] Hao Y, Bharathi MS, Wang L, Liu Y, Chen H, Nie S, et al. The role of surface oxygen in the growth of large single-crystal graphene on copper. Science. 2013;**342**:720-723. DOI: 10.1126/science.1243879

[53] Suzuki S, Terada Y, Yoshimura M. Suppression of graphene nucleation by turning off hydrogen supply just before atmospheric pressure chemical vapor deposition growth. Coatings. 2017;7:206. DOI: 10.3390/coatings7110206

[54] Chen M, Haddon R, Yan R, Bekyarova E. Advances in transferring chemical vapour deposition graphene: A review. Materials Horizons. 2017;**4**:1054-1063. DOI: 10.1039/C7MH00485K

[55] Lee HC, Liu WW, Chai SP, Mohamed AR, Aziz A, Khe CS, et al. Review of the synthesis, transfer, characterization and growth mechanisms of single and multilayer graphene. RSC Advances. 2017;7:15644-15693. DOI: 10.1039/C7RA00392G

[56] An CJ, Kim SJ, Choi HO, Kim DW, Jang SW, Jin ML, et al. Ultraclean transfer of CVD-grown graphene and its application to flexible organic photovoltaic cells. Journal of Materials Chemistry A. 2014;**2**:20474-20480. DOI: 10.1039/C4TA03432E

[57] Lin WH, Chen TH, Chang JK, Taur JI, Lo YY, Lee WL, et al. A direct and polymer-free method for transferring graphene grown by chemical vapor deposition to any substrate. ACS Nano. 2014;**8**:1784-1791. DOI: 10.1021/nn406170d

[58] Chandrashekar BN, Deng B, Smitha AS, Chen Y, Tan C, Zhang H, et al. Roll-to-roll green transfer of CVD graphene onto plastic for a transparent and flexible triboelectric nanogenerator. Advanced Materials. 2015;**27**:5210-5216. DOI: 10.1002/adma.201502560

[59] Yang X, Peng H, Xie Q, Zhou Y, Liu Z. Clean and efficient transfer of CVD-grown graphene by electrochemical etching of metal substrate. Journal of Electroanalytical Chemistry. 2013;**688**:243-248. DOI: 10.1016/j.jelechem.2012.09.025

[60] Martins LG, Song Y, Zeng T, Dresselhaus MS, Kong J, Araujo PT. Direct transfer of graphene onto flexible substrates. Proceedings of the National Academy of Sciences. 2013;**110**:17762-17767. DOI: 10.1073/pnas.1306508110

Chapter 5

# A DFT Investigation on Different Graphene Based Substrates on SERS: A Case Study of $TiO_2$ Adsorbed Gold/Graphene

*Hosein Hasan Bouzari, Rasoul Malekfar and Laleh Farhang Matin*


## Abstract

In this study, the Raman and the surface-enhanced Raman scattering (SERS) spectra of $TiO_2$ adsorbed on gold/graphene cluster is explained by density functional theory (DFT) calculations. we concentrated on the interaction between $TiO_2$, Au, and graphene, in which graphene is presented as a substrate component in SERS. Results indicate that changing graphene type including pure graphene and B/N-doped graphene enables modifications of interaction between molecule, gold, and graphene cluster. The Raman and SERS spectra of compounds are compared considering the intensity of spectra, which demonstrate the intensity dependence on graphene type which is due to the effects of chemical and electromagnetic properties.


**Keywords:** $TiO_2$, DFT, gold, graphene, SERS

## 1. Introduction

Graphene-enhanced Raman scattering (GERS) and graphene- mediated SERS (G-SERS) have been known for molecular trace detection in the nanomaterial substrate surface [1, 2].

In SERS, the Raman scattering intensity is related to laser intensity $I_L$ and polarizability tensor $\alpha$ by the expression [3]:

$$I = \frac{8\pi(\omega \pm \omega_{I'I})^4 I_L \sum \alpha_{\sigma\rho}^2}{9C^4}$$

Where $\omega$ is the frequency of incident laser and $\omega_{I'I}$ is the molecular transition frequency between states I and I′.

Electromagnetic effect leads to amplification of laser and scattering intensity. Chemical effects, including static chemical enhancement, molecular resonance, and charge-transfer effect, contribute to the change in polarizability derivatives with respect to internal coordinates.



IntechOpen

Graphene is used as part of SERS substrates to prepare additional chemical enhancement also as an assistant for "hot spots, which effectively amend the spectral enhancement result. The Raman spectra amplification of the molecules that were attracted to the graphene substrate occurred due to charge transfer [4, 5].

So far, few studies based on quantum chemical calculations for GERS and G-SERS process have been presented, due to the great challenges in modeling the graphene-molecule structure (including cluster model and periodic model) and the related charge transfer effect during the SERS process [6].

For gold component, SERS mechanisms involve off-resonance chemical effect (static chemical enhancement), on-resonance chemical effect) molecular resonance and charge-transfer resonance (and electromagnetic effect (surface plasmon resonance). DFT calculations take into account the effect of molecule-metal interaction and changes in electronic structure for both off- and on-resonance chemical enhancement. Molecular resonance is not considered due to absence of excitation in visible region.

Titanium dioxides ($TiO_2$) due to its interesting general properties including photocatalysis, catalysis, antibacterial effects, and in civic as self-cleaning that affect the quality of life are taken into consideration [7, 8]. The attractive physical and chemical features of $TiO_2$ depend on the crystal phase, size, and form of particles. For example, varying phases of crystalline $TiO_2$ have different band gaps of the rutile phase of 3.0 eV and anatase phase of 3.2 eV; defining the photocatalytic efficiency of $TiO_2$ [9].

Titanium dioxide nanoparticles with a high refractive index ($n = 2.4$) are appropriate for toothpaste, pharmaceuticals, coatings, papers, inks, plastics, food products, cosmetics, and the textile industry [10].

One of the best materials used in the above cases is $TiO_2$ doped with noble metals such as gold [11]. $TiO_2$ between B/N-doped graphene and Au clusters as electrodes, enhanced charge transport properties shows.

In this work, we concentrated on the interaction between $TiO_2$, Au, and graphene, in which graphene is presented as a substrate component in SERS.

For investigation of the chemical enhancement mechanism related to graphene - based SERS, we used density functional theory (DFT) on the $TiO_2$ /gold complexes with pure graphene, and B/N doped graphene.

## 2. Results and discussion

Gaussian 09 packages were applied for Theoretical calculations and Hyperchem software for drawn molecules. By B3LYP functional, Molecules and compounds were optimized and calculated. The basis sets for graphene, B/N doped graphene were described using basis sets of 3–21 + G (d, p) and the basis sets for Au atoms were used LANL2DZ.

Excitation spectra were calculated using DFT-optimized configurations with the CAM-B3LYP (long-range corrected effect) functional and the same basis sets [12, 13].

As shown in **Figure 1**, geometric optimization was applied on different compounds including $TiO_2$/Au-graphene, and $TiO_2$/Au-graphene-Au with complete graphene cluster, $TiO_2$/graphene (internal), $TiO_2$/Au-$G_N$ (N-doped graphene), and $TiO_2$/Au-$G_B$ (B-doped graphene).

Calculated Raman spectra of $TiO_2$/gold/graphene compounds were shown in **Figures 2** and **3**. The (a) in **Figure 2** corresponds to the Raman spectrum of $TiO_2$

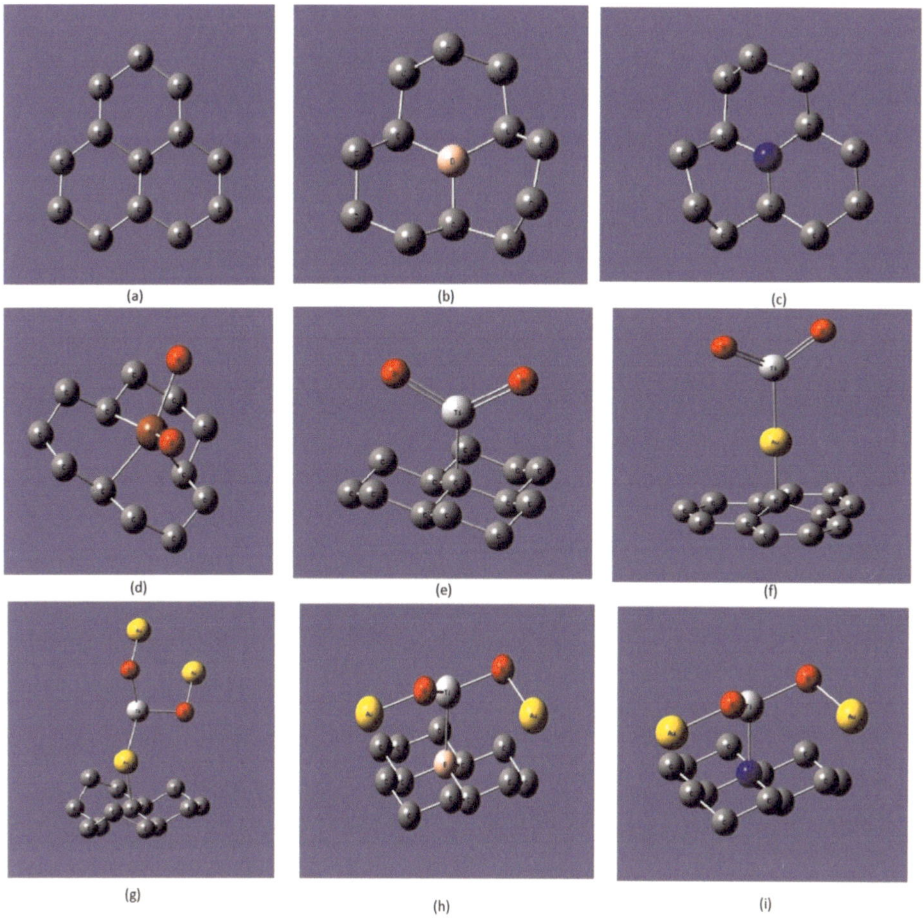


**Figure 1.**
*Geometric optimization of $TiO_2$/gold/graphene compounds: (a) pure graphene (b) B-doped graphene (c) N-doped graphene (d) $TiO_2$/ graphene (internal) (e) $TiO_2$/graphene (external) (f) $TiO_2$/Au-graphene (g) $TiO_2$/3Au-graphene (h) $TiO_2$/Au-$G_B$ (B-doped graphene) (i) $TiO_2$/Au-$G_N$ (N-doped graphene).*

molecule, the (b) in **Figure 2** is the calculated Raman spectrum of the $TiO_2$/Au-G complex, and the (c) in **Figure 2** is the calculated Raman spectrum of the $TiO_2$/Au-G-Au complex based on pure graphene cluster.

The major characteristic Raman peaks of $TiO_2$ molecule are located at 420 $cm^{-1}$, 490 $cm^{-1}$, 650 $cm^{-1}$, and 760 $cm^{-1}$. In the Raman spectrum of the TiO2/Au- graphene composite, the major characteristic Raman peaks at 490 $cm^{-1}$ and 760 $cm^{-1}$ red-shifted to 470 $cm^{-1}$ and 755 $cm^{-1}$, respectively. In the Raman spectrum of the TiO2/Au- graphene -Au composite, the major characteristic Raman peaks at 470 $cm^{-1}$ and 755 $cm^{-1}$ red-shifted to 440 $cm^{-1}$ and 750 $cm^{-1}$, respectively. The red-shift of Raman peaks is related to changes in the linked bond distances. For instance, longer Ti-O bond distance caused by Ti-Au binding leads to smaller wavenumber (lower force constants) of Ti-O stretching vibration mode.

We have compared the Raman intensities of $TiO_2$/gold/graphene compounds. As shown in **Figure 2**, we can see that the Raman intensity of $TiO_2$/Au- graphene -Au is larger than that of $TiO_2$ molecule. For Raman peaks position at 490 $cm^{-1}$and 760 $cm^{-1}$

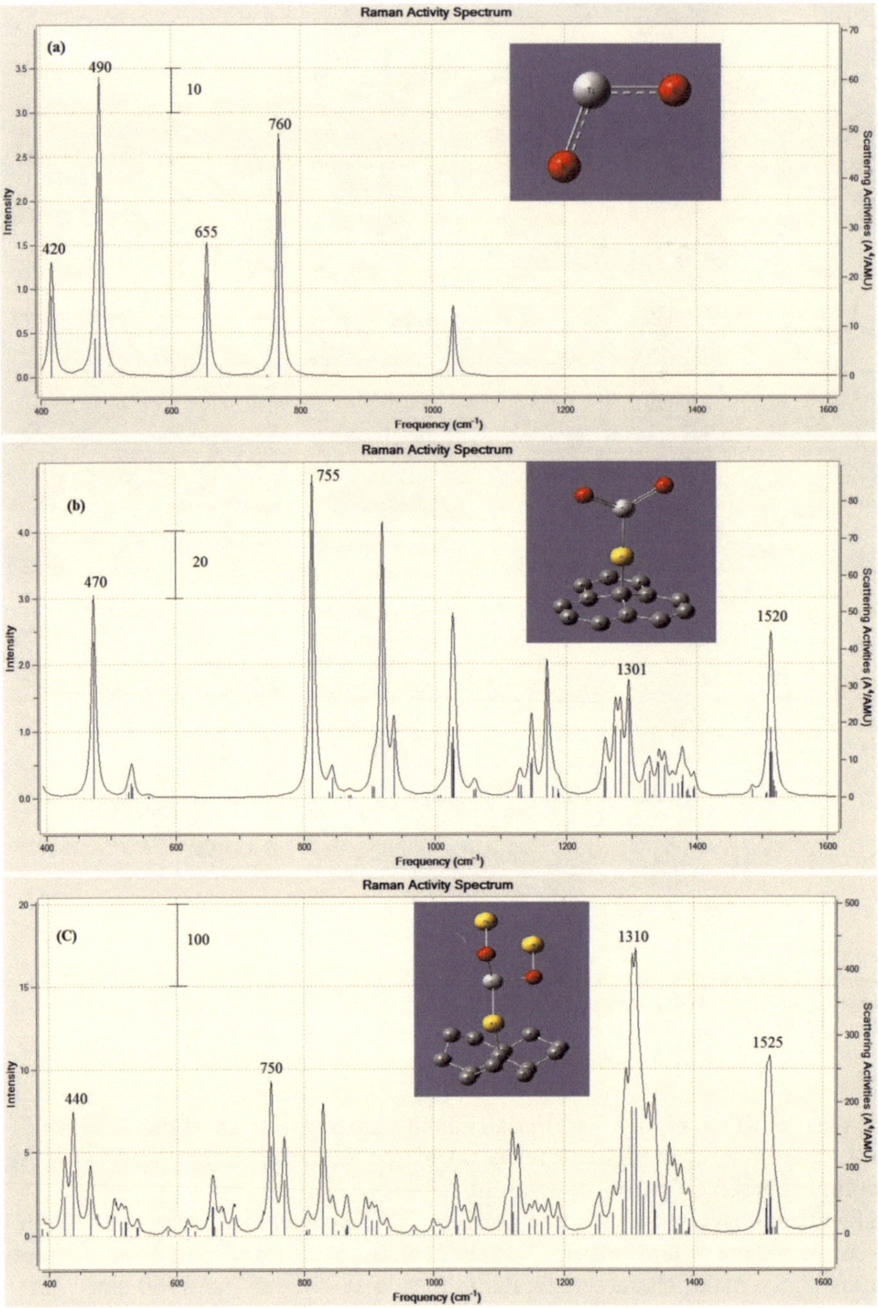


**Figure 2.**
*Raman spectra of $TiO_2$/gold/graphene compounds, (a) $TiO_2$ molecule (b) $TiO_2$/Au-graphene (c) $TiO_2$/Au-graphene-Au.*

($TiO_2$), the Raman intensities are about 60 (a.u.) and 50 (a.u.), respectively and for $TiO_2$/Au- graphene -Au are about 200 (a.u.) and 220 (a.u.), respectively. The reason can be considered due to the intensity enhancement by Au atoms.

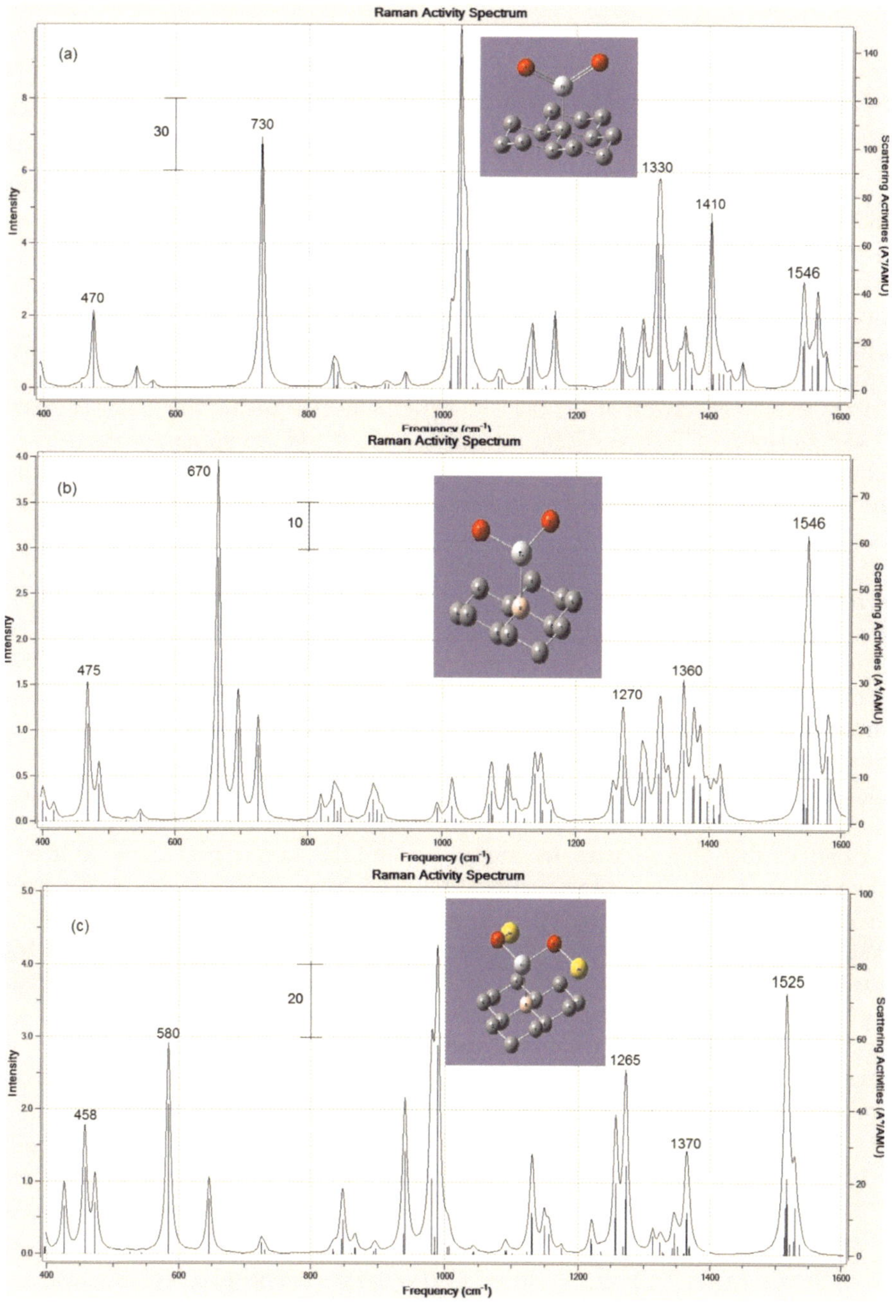


**Figure 3.**
*Raman spectra of (a) $TiO_2$/graphene (b) $TiO_2/G_B$ (c) $TiO_2/G_B$-Au.*

The SERS enhancement factor (EF) has been defined: $EF = \frac{I_{SERS}}{I_{Normal}}$ where $I_{SERS}$ and $I_{Normal}$ are the peak intensity on $TiO_2$/Au- graphene -Au and $TiO_2$, respectively. For peaks position at 490 $cm^{-1}$and 760 $cm^{-1}$, the EF is: [14]

$$(EF)_{490} = \frac{I_{SERS}}{I_{Normal}} = \frac{200}{60} = 3.3$$

$$(EF)_{760} = \frac{I_{SERS}}{I_{Normal}} = \frac{220}{50} = 4.4$$

The intensity of the SERS signals is enhanced by up to 3 fold. The Raman spectra amplification of the molecules can be related to the due to charge transfer between $TiO_2$ molecule and Au atoms.

The (a) in **Figure 3** corresponds to the Raman spectrum of $TiO_2$/G complex, the (b) in **Figure 3** is the calculated Raman spectrum of the $TiO_2$/ $G_B$ (boron doped) complex, and the (c) in **Figure 3** is the calculated Raman spectrum of the $TiO_2$/$G_B$-Au complex.

In the Raman spectrum of (a) the $TiO_2$/graphene composite, the major characteristic Raman peaks of $TiO_2$ at 470 $cm^{-1}$ and 730 $cm^{-1}$ and the Raman peaks of graphene at 1330 $cm^{-1}$, 1410 $cm^{-1}$ and 1546 $cm^{-1}$. In the Raman spectrum of (b) the $TiO_2$/ $G_B$ composite, the major characteristic Raman peaks red-shifted 470 $cm^{-1}$, 670 $cm^{-1}$, 1270 $cm^{-1}$, 1360 $cm^{-1}$ and 1546 $cm^{-1}$, respectively. In the Raman spectrum of (c) the $TiO_2$/$G_B$-Au composite, the major characteristic Raman peaks red-shifted 458 $cm^{-1}$, 580 $cm^{-1}$, 1265 $cm^{-1}$, 1370 $cm^{-1}$ and 1525 $cm^{-1}$, respectively. The red-shift of Raman peaks is related to changes in the linked bond distances.

The intensities of all particular Raman peaks were significantly improved under the Presence of gold atoms (**Figure 3c**). Whereas, with the presence of boron atom, the intensity of the Raman peaks decreases and the perturbations increases (**Figure 3b**), which show the presence of impurities in the graphene.

As shown in **Figure 4**, we have compared the Raman intensities of $TiO_2$/ $G_B$ and $TiO_2$/ $G_N$ compounds. We can see that the Raman intensity of $TiO_2$/ $G_N$ (**Figure 4b**) is lower than of $TiO_2$/ $G_B$ (**Figure 4a**) composite and the Raman peaks were shifted to smaller wavenumber (lower force constants). This could be due to the weaker bands of $TiO_2$/ $G_N$ composite than the $TiO_2$/$G_B$ composite.

We have compared the HOMO (highest occupied molecular orbital) and LUMO (lowest unoccupied molecular orbital) of TiO2/gold/graphene compounds. The energy difference between the HOMO and LUMO is named the HOMO–LUMO gap [15]. The difference in energy between these two frontier orbitals can be used to estimate the strength and stability of complexes [16]. The higher of the gap's energy, the molecule will be more stable and the harder to excite.

As shown in **Figure 5**, we have compared the HOMO and LUMO of $TiO_2$/ $G_B$, $TiO_2$/ $G_N$ $TiO_2$/ $G_B$-Au and $TiO_2$/$G_N$-Au composites.

$$(\Delta E)_{TiO_2/G_B} = -0.03858 - (-0.17867) = 0.14009\ Hartree$$

$$(\Delta E)_{TiO_2/G_N} = 0.01502 - (-0.11311) = 0.12813\ Hartree$$

$$(\Delta E)_{TiO_2/G_B-Au} = -0.08651 - (-0.24520) = 0.15869\ Hartree$$

$$(\Delta E)_{TiO_2/G_N-Au} = -0.08106 - (-0.22632) = 0.14526\ Hartree$$

The HOMO and LUMO energy of the compounds is calculated above. The largest energy difference is related to the composite of TiO2/$G_B$-Au with the value of 0.15869 Hartree.

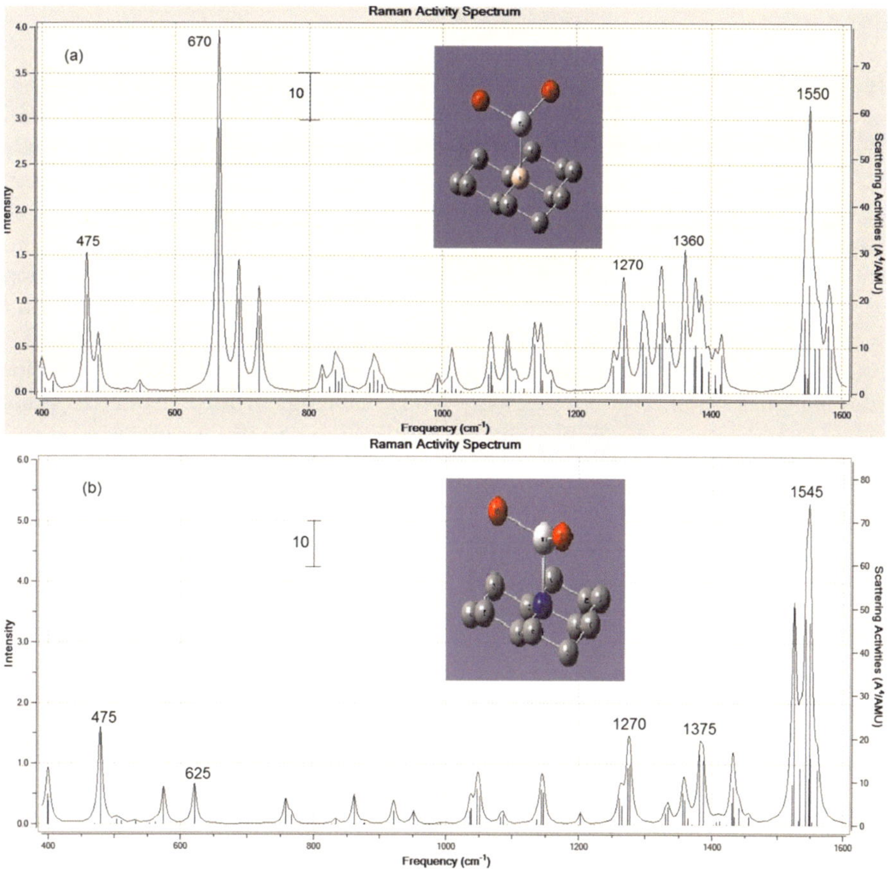


**Figure 4.**
*Raman spectra of (a) $TiO_2$/ $G_B$ (b) $TiO_2$/$G_N$.*

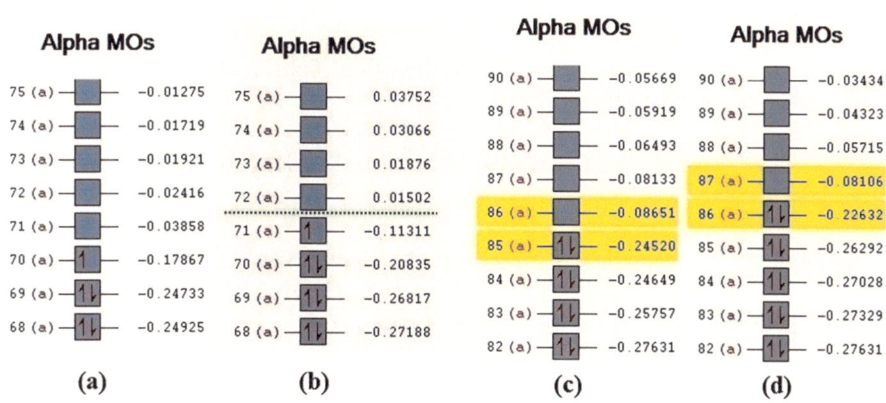


**Figure 5.**
*The HOMO and LUMO of (a) TiO2/ $G_B$ (b) TiO2/ $G_N$ (c) $TiO_2$/ $G_B$-Au (d) TiO2/$G_N$-Au composites.*

## 3. Conclusion

In summary, a DFT study is introduced on the SERS of TiO2 adsorbed Gold/Graphene by different Graphene based Substrates. Different types of graphene substrate cluster (perfect graphene, graphene with B- and N-doped graphene) were included to check the influence of graphene surface on spectral and energy properties. TiO2/ $G_B$-Au complex exhibited the largest chemical enhancement in calculated static Raman spectrum. The excitation properties including the transition states and pre-resonance Raman spectra were calculated, which shows the effect of the transition state on the resonance chemical enhancement as well as the relationship between the excitation energy and the non-resonant chemical enhancement. The HOMO and LUMO energy of TiO2/$G_B$-Au complex was the largest energy (0.15869) which indicated the greater stability of the molecule. The knowledge profited from this study is expected to be help to both graphenebased SERS study and ultrasensitive spectroscopic techniques.

## Author details

Hosein Hasan Bouzari[1*], Rasoul Malekfar[2] and Laleh Farhang Matin[3]

1 Department of Health and Biomedical Engineering, Tehran Medical Sciences, Islamic Azad University, Tehran, Iran

2 Department of Physics, Faculty of Basic Sciences, Tarbiat Modares University, Tehran, Iran

3 Department of Physics, North Tehran Branch, Islamic Azad University, Tehran, Iran

*Address all correspondence to: hh.bouzari@yahoo.com

## References

[1] Fleischmann M, Hendra PJ, McQuillan AJ. Raman spectra of pyridine adsorbed at a silver electrode. Chemical Physics Letters. 1974;**26**:163

[2] Kang L, Chu J, Zhao H, Xu P, Sun M. Recent progress in the applications of graphene in surface-enhanced Raman scattering and plasmon-induced catalytic reactions. Journal of Materials Chemistry C. 2015;**3**:9024

[3] Lombardi JR, Birke RL, Lu T, Xu J. Charge-transfer theory of surface enhanced Raman spectroscopy: Herzberg–Teller contributions. The Journal of Chemical Physics. 1986;**84**: 4174

[4] Xu H, Bjerneld EJ, Käll M, Börjesson L. Spectroscopy of single hemoglobin molecules by surface enhanced Raman scattering. Physical Review Letters. 1999;**83**:4357

[5] Xu H, Aizpurua J, Käll M, Apell P. Electromagnetic contributions to single-molecule sensitivity in surface-enhanced Raman scattering. Physical Review E. 2000;**62**:4318

[6] Zhao X, Chen M. Charge transfer mechanism of SERS for metal–molecule–metal junction supported by graphene and boron-doped graphene. RSC Advances. 2014;**4**:63596

[7] Li R, Li T, Zhou Q. Impact of titanium dioxide ($TiO_2$) modification on its application to pollution treatment—A review. Catalysts. 2020;**10**:804

[8] Hanaor DAH, Sorrell CC. Sand supported mixed-phase TiO2 photocatalysts for water decontamination applications. Advanced Engineering Materials. 2014;**16**(2): 248-254

[9] Liu L, Zhao H, Andino JM, Li Y. Photocatalytic CO2 reduction with H2O on TiO2 nanocrystals: Comparison of anatase, rutile and brookite polymorphs and exploration of surface chemistry. ACS Catalysis. 2012;**2**:1817-1828

[10] Vijayalakshmi R, V. Rajendran synthesis and characterization of nano-TiO2 via different methods. Archives of Applied Science Research. 2012;**4**(2): 1183-1190

[11] Begum T, Gogoi PK, Bora U. Photocatalytic degradation of crystal violet dye on the surface of Au doped TiO2 nanoparticle. Indian Journal of Chemical Technology. 2017;**24**:97-101

[12] Li N, Luo Q. Theoretical Chemistry Accounts. 2011;**130**:1023

[13] Lu T, Chen F. Journal of Computational Chemistry. 2012;**33**:580

[14] Willets K, Mayer K. Handbook of Organic Materials for Electronic and Photonic Devices. 2nd ed. Elsevier; 2019

[15] Griffith JS, Orgel LE. Ligand field theory. Quarterly Reviews, Chemical Society. 1957;**11**:381-383

[16] Kitagawa Y, Asaoka M, Natori Y, Miyagi K, Teramoto R, Matsui T, et al. Theoretical study on relationship between spin structure and electron conductivity of one-dimensional tri-nickel(II) complex. Polyhedron. 2017;**136**:125-131

Chapter 6

# Structural, Electronic, and Optical Properties of Mono- and Co-Doped Graphene with Ti and Ru

*Lutendo Phuthu, Nnditshedzeni Eric Maluta and Rapela Regina Maphanga*

## Abstract

Due to its properties, graphene is considered a revolutionary material for the future, and as a two-dimensional material it has received a lot of research attention over the last two decades. For graphene to be used in different technologies such as solar cells, much more work needs to be done to understand its properties and engineer its properties by combining it with other materials such as semiconductors. This research work reports computational investigation of the electronic and optical properties of Ti and Ru mono-doped and co-doped graphene. Geometry optimizations for the electronic and optical properties were performed by first-principles calculations based on density functional theory. Various supercells of graphene were modeled and optimized, and their properties were calculated. The results show that different graphene supercells have different electronic and optical properties. The energy bandgap of pure graphene is zero, and after doping with Ti and Ru it increases to 0.550 eV, and 0.786 eV, respectively. The co-doped graphene bandgap is 0.272 eV. The calculated optical properties showed that doping graphene with Ti and Ru shifts the absorption from the visible to the near-infrared region, and these results open possibilities of using doped graphene as a semiconductor material.

**Keywords:** graphene, density functional theory, bandgap, doping, optical properties

## 1. Introduction

Graphene has sparked great interest in recent decades due to its remarkable electrical and optical capabilities, as demonstrated by a groundbreaking experiment in graphene research in 2004 [1]. Graphene is a honeycomb-shaped two-dimensional sheet crystalline structure of atomically thick sp2-hybridized carbon (each carbon fortifies covalently with three other carbon atoms) [2–5]. It serves as a building block for various carbon dimensionalities, such as zero-dimensional Buckyball, one-dimensional nanotube, and three-dimensional graphite [2]. A pristine graphene has zero bandgap, because its conduction and valence bands meet at a single location at the Dirac points [6–12]. Graphene is considerably stable due to the tight packing of carbon atoms and hybridization of sp2, but only when the graphene size is smaller than 20 nm; otherwise, it is thermodynamically unstable [13, 14]. The classification of graphene as a metal, nonmetal, or semimetal is still up for discussion [13].

The properties of graphene depend on the number of graphene layers. A pristine graphene, for example, has a theoretical surface area of 2630 $m^2g^{-1}$, which is more than the surface area of carbon nanotubes (100–1000 $m^2g^{-1}$) [15, 16]. Furthermore, as compared to graphene with a few layers, a single-layer graphene has a higher surface area [14]. According to numerous studies, graphene has a high charge carrier mobility of 250,000 $cm^2v^{-2}\ s^{-1}$ at room temperature [5, 17, 18]. Furthermore, each layer of graphene absorbs up to 2.3% of the incident light with a reflectance of less than 0.1% [6]. As a result, it has a very high optical transparency of 97.7% as well as a high degree of flexibility [6, 19]. At room temperature, a single-layer graphene has a high thermal conductivity of 3000–5000 $Wm^{-1}\ K^{-1}$ [20]. Other properties include an electrical conductivity of 6000 S $cm^{-1}$ [21] and a Young's modulus of 1.0 TPa [22].

Graphene offers potential application in areas such as high-speed electronics, data storage, liquid crystal display (LCD) smart windows, organic light emitting diode (OLED), supercapacitors, solar cells, and electrochemical sensing [19, 23]. The combination of high electrical conductivity, chemical and thermal stability, and outstanding stretchability provides significant benefits for employing graphene as a transparent conductor in organic electronic devices. It is mostly used as a hybrid with other materials to enhance the properties of other materials making them stronger, valuable, and light weight [24–28]. Studies have shown that the number of graphene layers, defects in graphene layers, various concentrations of graphene, and different sizes of graphene have impact on properties of graphene [28–32]. Graphene has been modified in various ways to broaden its application in a variety of fields. One method to modify graphene is to introduce foreign elements into it to tempt its electronic properties. Mukherjee and Kaloni investigated the effect of boron and nitrogen doping on graphene. Their calculations showed that N-doped graphene had a Dirac point shift below the Fermi level and B-doped graphene had a Dirac point shift above the Fermi level, resulting in a bandgap opening. The opening of the bandgap appears at the Fermi level for co-doped graphene [33]. Sara Varghese *et al.* investigated the structural, energetic, and electronic properties of graphene doped with boron and nitrogen atoms at different doping concentrations. They observed that doping increases the bandgap and decreases the energetic stability [34]. Olaniyan *et al.* conducted a systematic study of the stability, electronic, and optical properties of mono- and co-doped graphene with beryllium and nitrogen. Be-N was found to be more stable than Be-doped graphene. The study also shows that when graphene is doped with Be and N, it transforms from semi-metallicity to semi-conductivity [35]. Despite the substantial amount of work that has been put into the theory and experimentation of doped graphene, there are still a great many applications that have not been fulfilled. As a result, research into doped graphene systems with superior performance continues to be pushed forward.

First-principle calculations are used in this study to evaluate several graphene supercells and examine the effects of those supercells on the electrical and optical properties of graphene material. In addition, it investigates how the electrical and optical properties of graphene change when doped with titanium or ruthenium and a combination thereof.

## 2. Computational details

Geometry optimizations for the electronic and optical properties were performed by the first-principle calculations based on DFT implemented in the Material Studio

CASTEP code, using the generalized gradient approximation (GGA) of Perdew-Burke-Ernzerhof (PBE), norm-conserving pseudopotential, periodic boundary conditions, and space group of P6/mmm. The k point was set at 6x6x1 with a cutoff energy of 350 eV and energy tolerance of 1.0x10–6 eV. The force tolerance was set at 0.03 eV, the displacement tolerance 0.0001 Å, and convergence threshold of $1.0 \times 10^{-6}$ eV/atoms.

## 3. Results and discussion

### 3.1 Structural properties of graphene supercells

A graphene unit cell of two carbons was constructed with a 1.42 Å C-C bond lengths, 120 bond angles, and the lattice parameters a = b = 2.46 Å and c = 6.8 Å (see **Figure 1a**). The unit cell was then extended to construct n × n supercells (where n is an integer number). A total of seven graphene supercells were constructed. A 4x4 supercell is shown in **Figure 1b**. A space group of P6/mmm was used for all supercells. The constructed supercells are listed in **Table 1**.

### 3.2 Electronic properties of graphene supercells

**Table 1** shows all the possible combinations of $n$ x $n$ graphene supercells constructed in this work along with the number of carbons of that supercell. We use C# to denote the number of carbons of an $n \times n$ supercell, where # is an integer number representing the number of carbons. The calculated band structures are shown in **Figures 2** and **3**. The Fermi level was set to zero and is indicated by the red dashed lines.

The band structures of the supercells studied show Fermi levels at the Dirac points, showing a bandgap of zero, which agrees with other literature [6–12]. It is observed

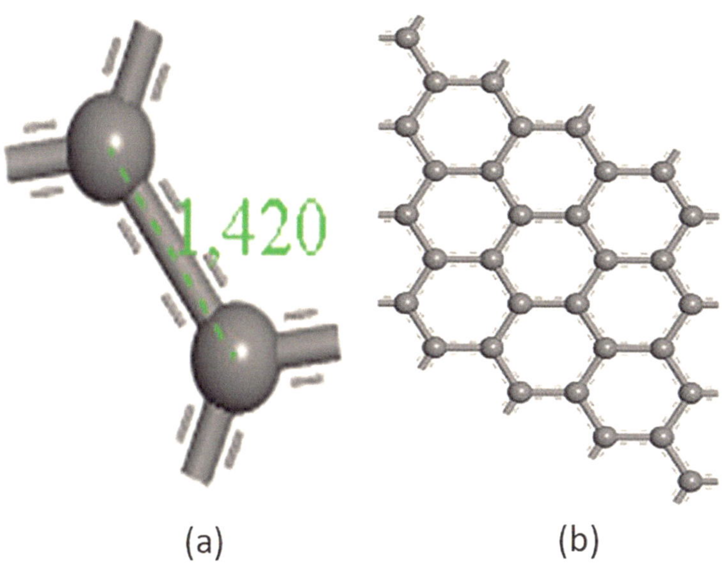


**Figure 1.**
*A four-atom unit cell of graphene.*

| Supercell ($n \times$ n) | Number of carbons |
|---|---|
| Unit cell | 2 (C2) |
| 2 × 2 | 8 (C8) |
| 3 × 3 | 18 (C18) |
| 4 × 4 | 32 (C32) |
| 5 × 5 | 50 (C50) |
| 6 × 6 | 72 (C72) |
| 7 × 7 | 98 (C98) |

**Table 1.**
*Different graphene supercells.*

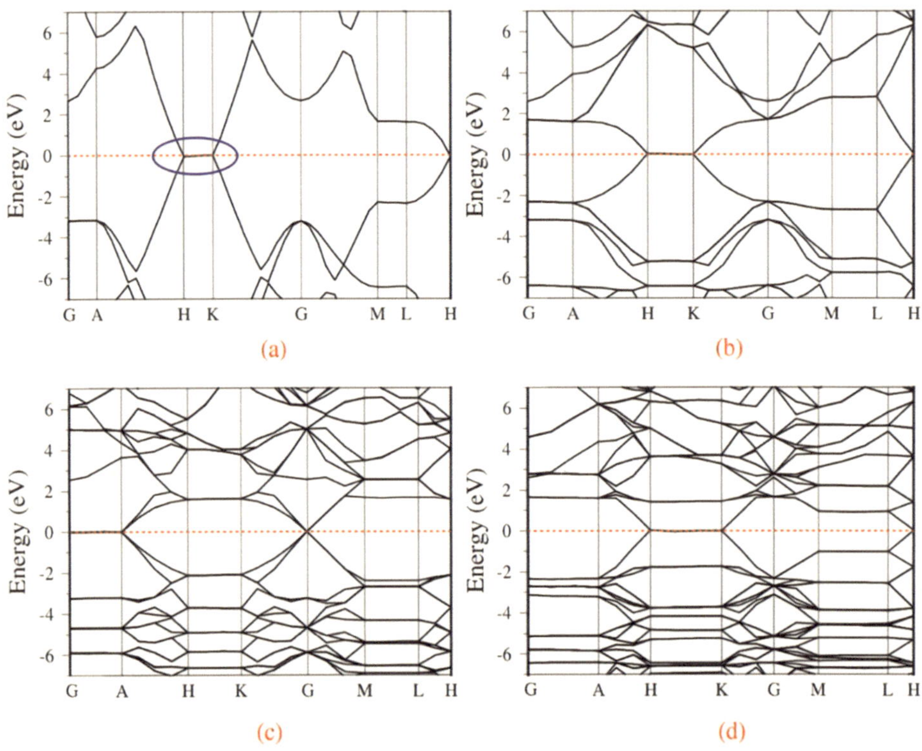


**Figure 2.**
*Calculated band structures of (a) C2, (b) C8, (c) C18 and (d) C32.*

that the bandgap energy of graphene is not affected when the size of the supercells is changed.

To investigate the nature of the states that comprise the conduction and valence band edges, we calculated the contributions of all atomic orbitals in the total density of states (TDOS) and the unique atomic shells in the partial density of states (PDOS) band edges. **Figures 4** and **5** show the density of states (DOSs), which describes the number of states per energy interval. The DOS agrees with the calculated band structure. At the Fermi level, the DOSs are very low, which are consistent with the calculated band structure.

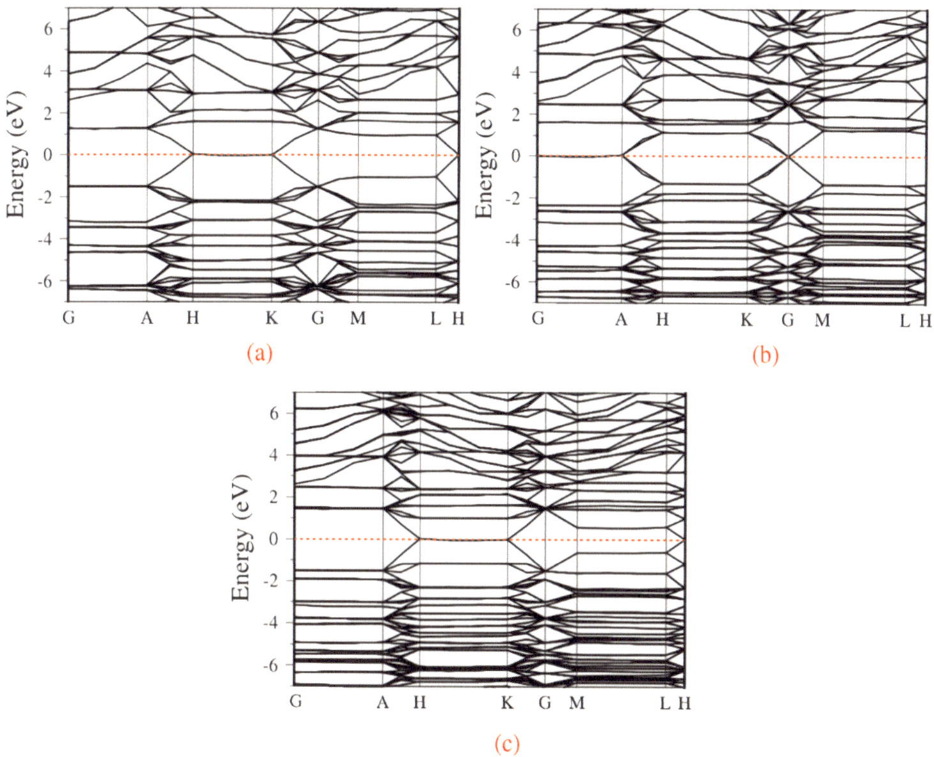


**Figure 3.**
*Calculated band structures of (a) C50, (b) C72, and (c) C98.*

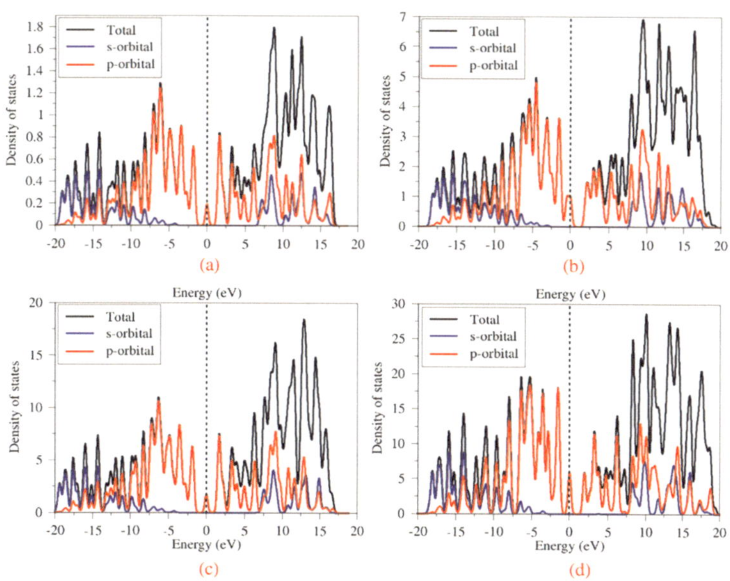


**Figure 4.**
*TDOS and PDOS for the (a) C2, (b) C8, (c) C18, and (d) C32 graphene supercells.*

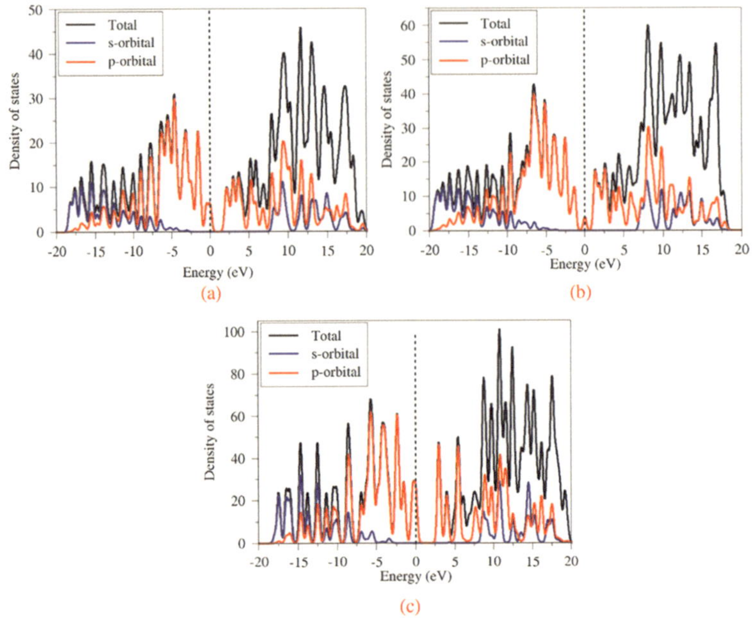


**Figure 5.**
*TDOS and PDOS for the (a) C50, (b) C72, and (c) C98 graphene supercells.*

It was found that the electron distribution in graphene is due to the contribution of the s and p atomic shells, which are responsible for the energy transfer in graphene. The results show that the s and p states are dominant in both the conduction and valence bands. However, at the Fermi level or near the Fermi level, only the p state is dominant.

## 3.3 Optical properties of graphene supercells

To investigate the optical response of graphene, we calculated its absorption, dielectric function, and refractive index. **Figure 6** illustrates the optical absorption calculations for different supercells. It can be observed that C8, C32, and C50 supercells exhibit strong absorption in the ultraviolet-visible range that extends into the infrared. The C2 and C18 supercells absorb more light in the UV area and dissipate in the visible region around the wavelength of 600 nm. C98 enhances the absorption activity to 700 nm, while C72 enhances it to 900 nm.

Dielectric materials tend to become polarized when exposed to an external electric field. The term "dielectric function" refers to the property of a substance that determines its polarization. The dielectric function is defined as follows:

$$\varepsilon = \varepsilon_1(\omega) + \varepsilon_2(\omega) \quad (1)$$

where $\varepsilon_1(\omega)$ and $\varepsilon_2(\omega)$ are the real and imaginary parts of the dielectric function. The real part of the dielectric function is connected to the material's polarization, whereas the imaginary part is related to the electronic absorption. **Figure 7** shows the calculated dielectric function of the seven graphene supercells up to a photon energy of 10 eV. In the limit of zero photon energy, the findings showed dielectric constant $\varepsilon_0$ values of 4.97, 6.82, 4.23, 11.92, 7.44, 6.07, and 6. 03 for the supercells C2, C8, C18, C32, C50, C72, and C98, respectively. The dielectric constant is proportional to the

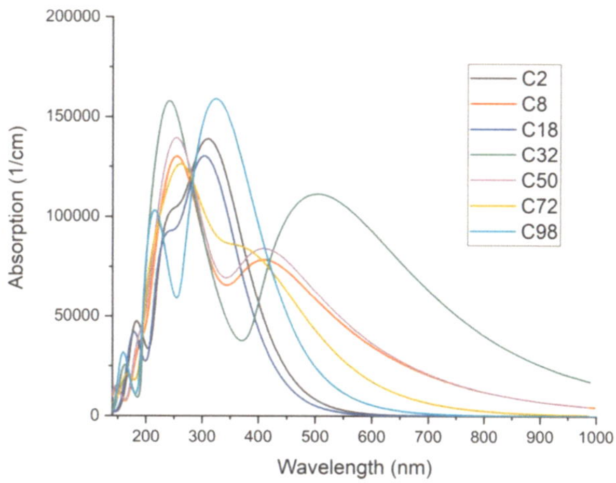


**Figure 6.**
*Calculated absorption properties of the graphene supercells.*

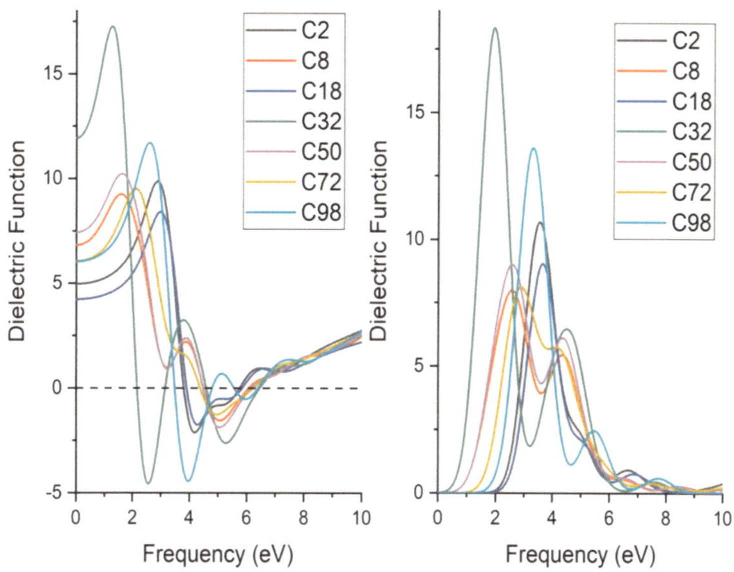


**Figure 7.**
*Calculated dielectric function properties of the graphene supercells.*

electric displacement, which is proportional to the polarization of the material. The imaginary part of the dielectric function shows that the low-frequency peaks are located at 1.93 to 3.62 eV.

The real part $n(\omega)$ and imaginary part $k(\omega)$ (extinction coefficient) of the refractive index are determined by the dielectric function using the Kramers-Kronig transformation and are defined as:

$$n(\omega) = \left(\frac{|\varepsilon(\omega)| + \varepsilon_1(\omega)}{2}\right)^{\frac{1}{2}} \tag{2}$$

$$k(\omega) = \left( \frac{|\varepsilon(\omega)| - \varepsilon_1(\omega)}{2} \right)^{\frac{1}{2}} \tag{3}$$

The extinction coefficient is relative to the amount of light absorbed. **Figure 8** shows the calculated refractive index of graphene supercells. The refractive index $n_0$ is equal to the $\sqrt{\varepsilon_0}$ in the limit of zero photon energy. These findings reveal refractive index $n_0$ values of 2.23, 2.61, 2.06, 3.45, 2.73, 2.46, and 2.46 for supercells C2, C8, C18, C32, C50, C72, and C98, respectively. The extinction coefficient has low-frequency peaks located from 2.28 to 3.97 eV.

### 3.4 Electronic properties of doped graphene

According to the findings of research conducted on the optical properties of various graphene supercells, the 4x4 graphene supercell demonstrates a superior optical response compared to other graphene supercells. These results have contributed to the author's decision to use the 4x4 graphene supercell and dope it with titanium and ruthenium atoms to study its electronic and optical properties. In the case of mono-doped graphene, the doping was accomplished by exchanging one of the graphene's carbon atoms for either titanium or ruthenium as shown in **Figure 9(a)**. On the other hand, in the case of co-doped graphene, two carbon atoms were exchanged for titanium and ruthenium, as illustrated in **Figure 9(b)**.

**Figure 10** presents an illustration of the band structure that was computed for the doped graphene 4x4 super cell. The results show that doping graphene causes an increase in the bandgap, which can be seen near the Fermi level in **Figure 10**. While the energy of the bandgap of Ti-doped graphene is 0.555 eV, the energy of the bandgap of Ru-doped graphene is 0.786 eV, and the energy of the bandgap of co-doped graphene is 0.272 eV. As a result of these properties, graphene can now be categorized as a material that falls into the category of semiconductor. The band

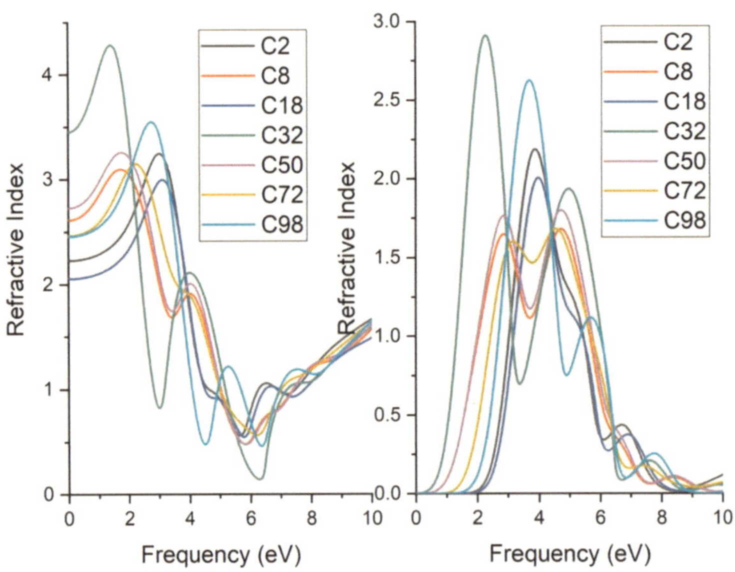


**Figure 8.**
*Calculated refractive index properties of the graphene supercells.*

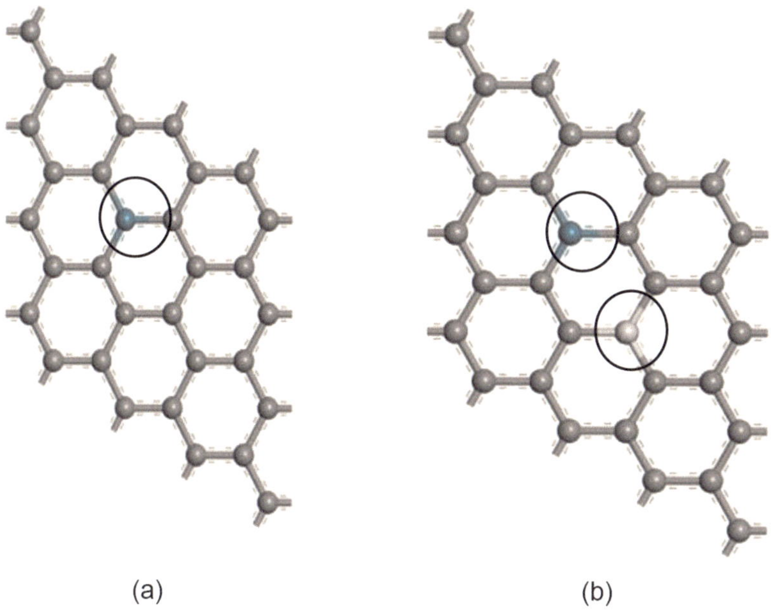


**Figure 9.**
*Structures of doped graphene: (a) mono-doped and (b) co-doped.*

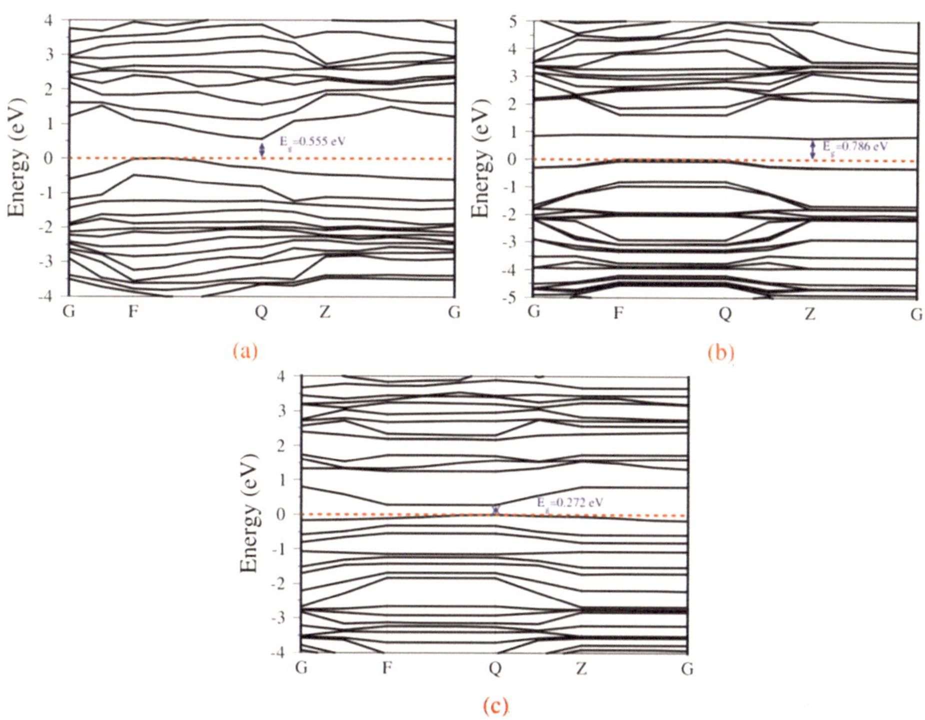


**Figure 10.**
*Calculated band structures of doped graphene with (a) Ti, (b) Ru, and (c) co-doped graphene.*

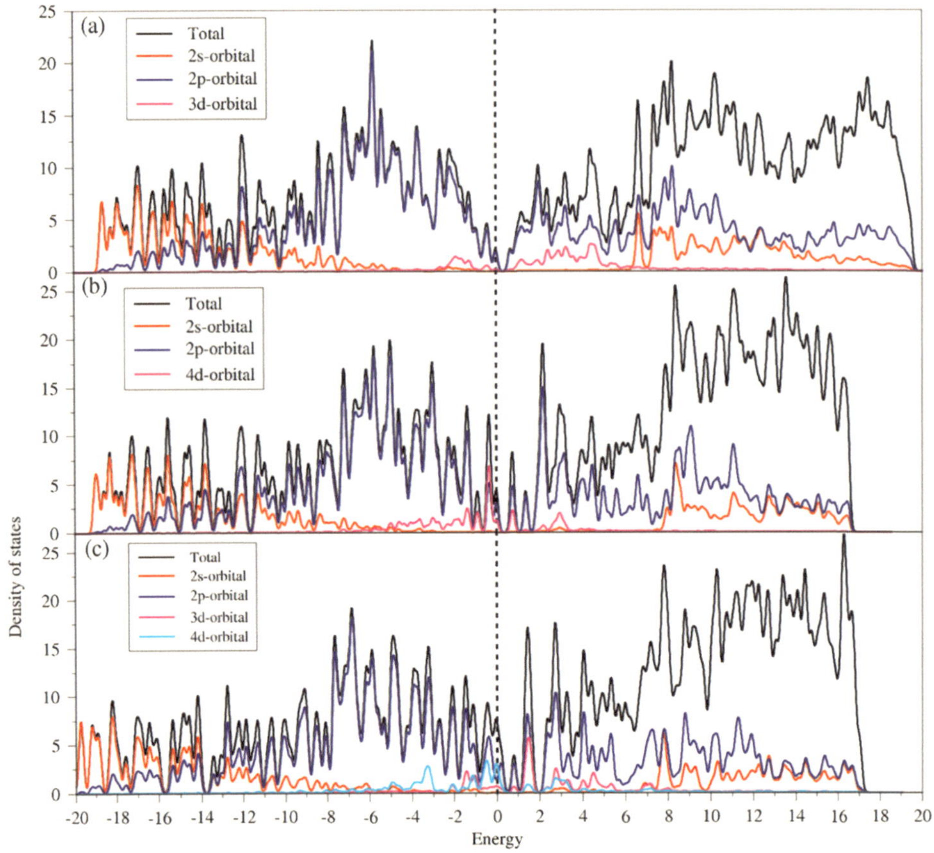


**Figure 11.**
*TDOS and PDOS for the (a) Ti-doped, (b) Ru-doped and co-doped graphene.*

structures shown in **Figure 10a** and **b** both exhibit an indirect bandgap, whereas the bandgap shown in **Figure 10c** exhibits a direct bandgap.

**Figure 11** shows the contribution made by doping graphene to the atomic shells. Doping graphene with the selected elements results in the introduction of numerous minor state density peaks, which can be clearly seen in the Fermi level region, as shown in **Figure 11**. This can be clearly observed by comparing **Figure 11** with the density of states of the 4x4 graphene supercell as shown in **Figure 5(d)**. Ru is a transition metal having a 4d electron configuration, whereas Ti has a 3d electron configuration. The 3d contribution from Ti is highlighted by the cyan color in **Figure 11a**, and the 4d contribution from Ru is highlighted by the cyan color in **Figure 11b**. Both of these contributions are in the energy range of −6 eV to 6 eV. In the vicinity of the Fermi level, the 3d state density has a modest peak, whereas the 4d state density has obvious peaks.

### 3.5 Optical absorption of mono and co-doped graphene

The calculated results of the optical absorbance of the doped graphene are illustrated in **Figure 12**. The results demonstrate that, when graphene is doped with titanium and ruthenium, there is no change in the activity of the absorption in the ultraviolet area. When Ti and Ru are added to graphene, a blue shift occurs in the

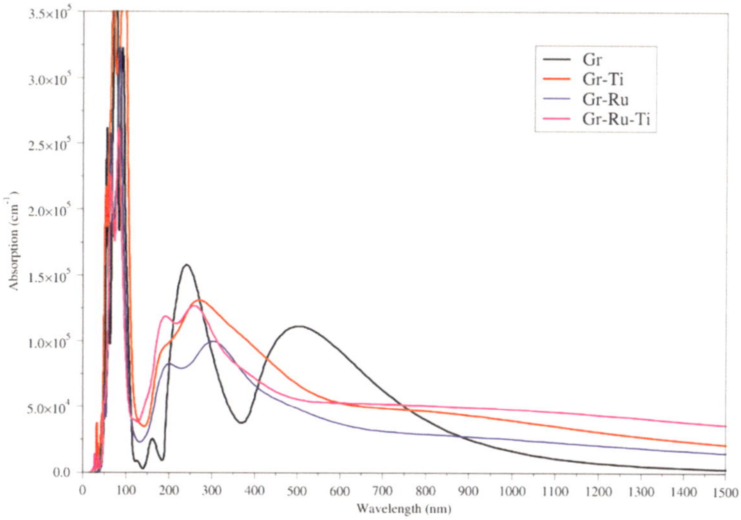


**Figure 12.**
*Calculated absorption properties for pristine and doped graphene.*

visible area. This blue shift is found for both mono-doped and co-doped graphene. Graphene with one dopant also exhibits this blue shift. Because of the dopants that were introduced into graphene, there is a red shift that occurs in the infrared region, which is located above 800 nm.

## 4. Conclusion

Using density functional theory as it is embodied in Material Studio, a number of electrical and optical properties were successfully explored. The results of our research indicate that the electrical structure of different graphene supercells is the same. The band structure computations showed that the Fermi level in all graphene supercells is located at the Dirac point, which indicates that there is no bandgap energy. According to the results of the density of state calculations, the s and p orbitals predominated in the valence band as well as the conduction band. According to the findings of the absorption, certain graphene supercells have minimal activity in the visible region, but other graphene structures have high absorbance in the visible area reaching all the way to the infrared sector of the spectrum. The findings suggest that graphene has a high dielectric constant, which qualifies it as an excellent candidate for application in various electrical devices. When graphene is mono-doped or co-doped with either titanium or ruthenium, a bandgap opening occurs. These results demonstrate a change from semi-metallic to semi-conducting behavior. This paves the way for the exploration of novel applications for graphene as a semiconductor.

## Acknowledgements

The authors acknowledge the Centre for High Performance Computing (CHPC) for computing.

Resources. LP thanks the National Institute for Theoretical and Computational Sciences (NITheCS) and Armaments Corporation of South Africa SOC Ltd. (ARMSCOR) for financial support.

## Author details

Lutendo Phuthu[1], Nnditshedzeni Eric Maluta[1,3*] and Rapela Regina Maphanga[2,3]

1 Department of Physics, University of Venda, Thohoyandou, South Africa

2 Next Generation Enterprises and Institutions Cluster, Council for Scientific and Industrial Research, Pretoria, South Africa

3 National Institute for Theoretical and Computational Sciences, (NITheCS), Gauteng, South Africa

*Address all correspondence to: eric.maluta@univen.ac.za

## References

[1] Novoselov KS, Geim AK, Morozov SV, Jiang D, Zhang Y, Dubonos SV, et al. Electric field effect in atomically thin carbon films. Science. 2004;**306**:666-669. DOI: 10.1126/science.1102896

[2] Geim AK, Novoselov KS. The rise of graphene. Nature Materials. 2007;**6**: 183-191. DOI: 10.1038/nmat1849

[3] Rao CN, Sood AK, Subrahmanyam KS, Govindaraj A. Graphene: The new two-dimensional nanomaterial. Angewandte Chemie International Edition. 2009;**48**: 7752-7777. DOI: 10.1002/anie.200901678

[4] Allen MJ, Tung VC, Kaner RB. Honeycomb carbon: A review of graphene. Chemical Reviews. 2009;**110**: 132-145. DOI: 10.1021/cr900070d

[5] Soldano C, Mahmood A, Dujardin E. Production, properties and potential of graphene. Carbon. 2010;**48**:2127-2150. DOI: doi.org/10.1016/j.carbon.2010.01.058

[6] Nair R, Blake P, Grigorenko A, Novoselov K, Booth T, Stauber T, et al. Fine structure constant defines visual transparency of graphene. Science. 2008; **320**:1308. DOI: 10.1126/science.115696

[7] Tiwari SK, Kumar V, Huczko A, Oraon R, Adhikari AD, Nayak G. Magical allotropes of carbon: Prospects and applications. Critical Reviews in Solid State and Materials Sciences. 2016; **41**(4):257-317. DOI: 10.1080/10408436.2015.1127206

[8] Chua CK, Pumera M. Chemical reduction of graphene oxide: A synthetic chemistry viewpoint. Chemical Society Reviews. 2014;**43**(1):291-312. DOI: doi.org/10.1039/C3CS60303B

[9] Su Y, Kravets V, Wong S, Waters J, Geim A, Nair R. Impermeable barrier films and protective coating based on reduced graphene oxide. Nature Communications. 2014;**5**:4843. DOI: 10.1038/ncomms5843

[10] Toh SY, Loh KS, Kamarudin SK, Daud WRW. Graphene production via electrochemical reduction of graphene oxide: Synthesis and characterisation. Chemical Engineering Journal. 2014;**251**: 422-434. DOI: doi.org/10.1016/j.cej.2014.04.004

[11] Poon J, Batchelor-McAuley C, Tschulic K, Compton RG. Single graphene nanoplacelets: Capacitance, potential of zero charge and diffusion coefficient. Chemical Science. 2015;**6**(5):2869-2876. DOI: doi.org/10.1039/C5SC00623F

[12] Bacon M, Bradley SJ, Nann T. Graphene quantum dots. Particle and Particle Systems Charactierization. 2014; **31**(4):415-428. DOI: doi.org/10.1002/ppsc.201300252

[13] Tiwari SK, Mishra RK, Ha SK, Huczko A. Evolution of graphene oxide and graphene: From imagination to industrialisation. Chemistry of Nanomaterials. 2018;**4**(7):598-620. DOI: doi.org/10.1002/cnma.201800089

[14] Tiwari SK, Sahoo S, Wang N, Huczko A. Graphene research and their outputs: Status and prospect. Journal of Science: Advanced Materials and Devices. 2020;**5**:10-29. DOI: doi.org/10.1016/j.jsamd.2020.01.006

[15] Zhu Y, Murali S, Cai W, Li X, Suk JW, Potts JR, et al. Graphene and graphene oxide: Synthesis, properties and applications. Advanced Materials. 2010;**22**:3906-3924. DOI: doi.org/10.1002/adma.201001068

[16] Weber CM, Eisele DM, Rabe JP, Liang Y, Feng X, Zhi L, et al. Graphene-based optically transparent electrodes for spectroelectrochemistry in the UV-vis region. Small. 2010;**6**:184-189. DOI: 10.1002/smll.200901448

[17] Ortila M, Faugeras C, Plochocka P, Neugebauer P, Martinez G, Maude DK, et al. Approaching the Dirac point in high mobility multilayer epitaxial graphene. Physical Review Letters. 2008; **101**:267601. DOI: 10.1103/PhysRevLett.101.267601

[18] Liao L, Peng H, Liu Z. Chemistry makes graphene beyond graphene. Journal of the American Chemical Society. 2014;**136**(35):12194-12200. DOI: doi.org/10.1021/ja5048297

[19] Pang S, Hernandez Y, Feng X, Müllen K. Graphene as transparent electrode material for organic electronics. Advanced Materials. 2011; **23**:2779-2795. DOI: doi.org/10.1002/adma.201100304

[20] Balandin AA, Ghosh S, Bao W, Calizo I, Teweldebrhan D, Miao F, et al. Superior thermal conductivity of single-layer graphene. Nano Letters. 2008;**8**: 902-907. DOI: doi.org/10.1021/nl0731872

[21] Du X, Skachko I, Barker A, Andrei EY. Approaching ballistic transport in suspended graphene. Nature Nanotechonology. 2008;**3**:491-495. DOI: 10.1038/nnano.2008.199

[22] Lee C, Wei X, Kysar JW, Hone J. Measurements of the elastic properties and intrinsic strength of monolayer graphene. Science. 2008;**321**:385-388. DOI: 10.1126/science.1157996

[23] Randviir EP, Brownson DAC, Banks CE. A decade of graphene research: Production, applications and outlook. Materials Today. 2014;**17**(9): 426-432. DOI: doi.org/10.1016/j.mattod.2014.06.001

[24] Dubey PK, Tripathi P, Tiwari RS, Sinha ASK, Srivastava ON. Synthesis of reduced graphene oxide–$TiO_2$ nanoparticle composite systems and its application in hydrogen production. International Journal of Hydrogen Energy. 2014;**39**:16282-16292. DOI: doi.org/10.1016/j.ijhydene.2014.03.104

[25] Xu P, Tang Q, Zhou Z. Structural and electronic properties of graphene-ZnO interfaces: Dispersion-corrected density functional theory investigations. Nanotechnology. 2013;**24**: 305401-305407. DOI: 10.1088/0957-4484/24/30/305401

[26] Martins PM, Ferreira CG, Silva AR, Magalhães B, Alves MM, Pereira L, et al. $TiO_2$/graphene and $TiO_2$/graphene oxide nanocomposites for photocatalytic applications: A computer modelling and experimental study. Composites Part B: Engineering. 2018;**145**:39-46. DOI: doi.org/10.1016/j.compositesb.2018.03.015

[27] An X, Yu JC. Graphene-based photocatalytic composites. RSC Advances. 2011;**1**:1426-1434. DOI: doi.org/10.1039/C1RA00382H

[28] Fang X, Li M, Guo K, Liu X, Zhu Y, Sebo B, et al. Graphene-compositing optimization of the properties of dye-sensitized solar cells. Solar Energy. 2014; **101**:176-181. DOI: doi.org/10.1016/j.solener.2013.12.005

[29] Ferrari AC, Meyer JC, Scardaci V, Casiraghi C, Lazzeri M, Mauri F, et al. Raman Spectrum of graphene and graphene layers. Physical Review Letters. 2006;**97**:187401-187404. DOI: doi.org/10.1103/PhysRevLett.97.187401

[30] Nilsson J, Neto AHC, Guinea F, Peres NMR. Electronic properties of

graphene multilayers. Physical Review Letters. 2006;**97**(26):266801-266804. DOI: 10.1103/PhysRevLett.97.266801

[31] Li X, Tao L, Chen Z, Fang H, Li X, Wang X, et al. Graphene and related two dimensional materials: Structure-property relationships for electronics and optoelectronics. Applied Physics Reviews. 2017;**4**:021306-021337. DOI: doi.org/10.1063/1.4983646

[32] Fan J, Liu S, Yu J. Enhanced photovoltaic performance of dye-sensitized solar cells based on $TiO_2$ nanosheets/graphene composite films. Journal of Materials Chemistry. 2012;**22**: 17027-17036. DOI: doi.org/10.1039/C2JM33104G

[33] Mukherjee S, Kaloni TP. Electronic properties of boron- and nitrogen-doped graphene: A first principles study. Journal of Nanoparticle Research. 2012; **14**:1059-1052. DOI: 10.1007/s11051-012-1059-2

[34] Varghese SS, Swaminathan S, Singh KK, Mittal V. Energetic stabilities, structural and electronic properties of monolayer fraphene doped with boron and nitrogen atoms. Electronics. 2016;**5**: 91. DOI: 10.3390/electronics5040091

[35] Olaniyan O, Maphasha RE, Madito MJ, Khaleed AA, Igumbor E, Manyala M. A systematic study of the stability, electronic and optical properties of beryllium and nitrogen co-doped graphene. Carbon. 2018;**129**: 207-227. DOI: 10.1016/j.carbon.2017.12.014

Chapter 7

# Anomalous Charge Transport Properties and Band Flattening in Graphene: A Quasi-Relativistic Tight-Binding Study of Pseudo-Majorana States

*Halina V. Grushevskaya and George Krylov*

## Abstract

Anomalous charge carrier transport in graphene is studied within a topologically nontrivial quasi-relativistic graphene model. The model predicts additional topological contributions, such as the Majorana-like mass-term correction to the ordinary ohmic component of the current, the spin-orbital-coupling, "Zitterbewegung"-effect corrections to conductivity in space, and time dispersion regime. The corrections appear due to non-Abelian quantum statistics for the charge carriers in graphene. The chiral anomaly of electrophysical and optical properties may emerge due to a deconfinement of the pseudo-Majorana quasiparticles. It has been shown that phenomena of negative differential conductivity, loss of universal far-infrared optical conductivity, and nonzero "minimal" direct-current conductivity in graphene occur due to flattening and vorticity of the pseudo-Majorana model graphene energy bands.

**Keywords:** graphene, anomalous charge transport, conductivity, chiral anomaly, quasi-relativistic model, Dirac-Hartree-Fock self-consistent field approximation, pseudo-Majorana fermion graphene model

## 1. Introduction

Graphene discovery is the driving force for progress in the current developments of quantum devices due to fascinating electrical and optical graphene properties. Graphene is an atomically thin carbon layer, in which valent and conduction bands touch each other in six points of hexagonal Brillouin zone, called Dirac points or valleys $K, K'$ [1]. Graphene belongs to strongly correlated systems, in which many-body interactions occur. Thanks to the strong electron-hole correlations, the electrophysical and optical properties of the graphene are very unusual. According to the linear response theory, the temperature dependence of the Hall conductivity for Fermi liquid, to which most metals belong, does not depend upon temperature [2]. But, the strongly correlated systems exhibit an anomalous charge magnetotransport

 

[3]. A similar situation prevails for universal nonzero direct-current (DC) graphene minimum conductivity (scale range of $\frac{4e^2}{h}$ [4] – $\frac{6e^2}{h}$ [5, 6] depending on support type) when the linear response theory predicts vanishing electrical conductivity for direct current (DC) or $\frac{4e^2}{\pi h}$ for the low-frequency conductivity in pure graphene [7, 8]. Here, $h$ is the Planck constant, and $e$ is the electron charge.

The situation with optical conductivity in the far-infrared spectral range of 0.05 – 0.5 eV (25 $\mu$m–350 $\mu$m) turned out to be even more unpredictable. Theoretical calculations of the far-infrared optical conductivity within a pseudo-Dirac massless fermion model for graphene predict its universal value, $\frac{e^2}{4\hbar}$, [9], which contradicts entirely experimental data on the existence of extrema in this spectral range, namely, an asymmetric peak at 0.15 eV ($\sim 150 - 200$ cm$^{-1}$) [10] and a shallow minimum in the spectral range of $\sim 0.2 - 0.3$ eV [11, 12] (see **Figure 1a** and **b**). Here, $\hbar$ is the Planck constant divided by $2\pi$. Slowly increasing the optical conductivity approaches the universal value in the mid-to-near-infrared spectral range of $0.5 - 1.2$ eV [13] where optical transitions occur far enough from the valleys $K, K'$ of the graphene Brillouin zone. Thus, the applicability of the pseudo-Dirac graphene model is doubtful in both spectral ranges.

The electron is a complex fermion; thus, if one decomposes its wavefunction into its real and imaginary parts, which would be Majorana fermions, they are rapidly remixed by electromagnetic interactions. However, such a decomposition could be reasonable for graphene because of the effective electrostatic screening. Experimental signatures of graphene Majorana states in graphene-superconductor junctions without the need for spin-orbital coupling (SOC) have been established in [14]. An analysis of Majorana-like graphene models also become relevant in connection with the discovery of unconventional superconductivity for twisted bilayer graphene at $\theta \approx 1.05$ angle of rotation of one monoatomic layer (monolayer) relative to another one [15]. A quantum statistics of graphene model with Majorana-like states should be a non-Abelian one, and the absence of that is the main obstacle to shed light on topologically nontrivial mechanisms of graphene conductivity. The non-Abelian pseudo-Majorana statistics of graphene charge carriers, which could clarify the anomalous effects, are behind several major theoretical approaches to graphene. The approaches are based on the massless pseudo-Dirac fermion model. The pseudo-Dirac graphene model Hamiltonian gives predictions that are very different from experimental data for a wide range of transport phenomena in graphene physics, such as the existence of topological currents in graphene superlattices [16], a sharp rise of Fermi velocity value $v_F$ in touching valent and conduction bands [17], and a lack of excitonic instability [18].

So, up to now, there is no consistent theory of topologically nontrivial graphene conductivity.

In this chapter, we show that charge transport coefficients in linear response to electromagnetic fields contain anomalous contributions arising from strong correlations between graphene charge carriers. The strong temperature dependencies that are observed in Hall, and optical and electrical conductivities are explained by non-Abelian statistics of topologically nontrivial graphene charge carriers of pseudo-Majorana nature within an earlier developed quasi-relativistic self-consistent Dirac-Hartree-Fock approach in a tight-binding approximation [19–25]. This quasi-relativistic approach allowed to achieve success in calculations of electronic properties for quasi-circular graphene $p - n(n - p)$ junctions also [26].

Our goal is to study quantum transport of charge carriers with vortex dynamics within a quasi-relativistic graphene model using a high-energy $\vec{k} \cdot \vec{p}$-Hamiltonian and

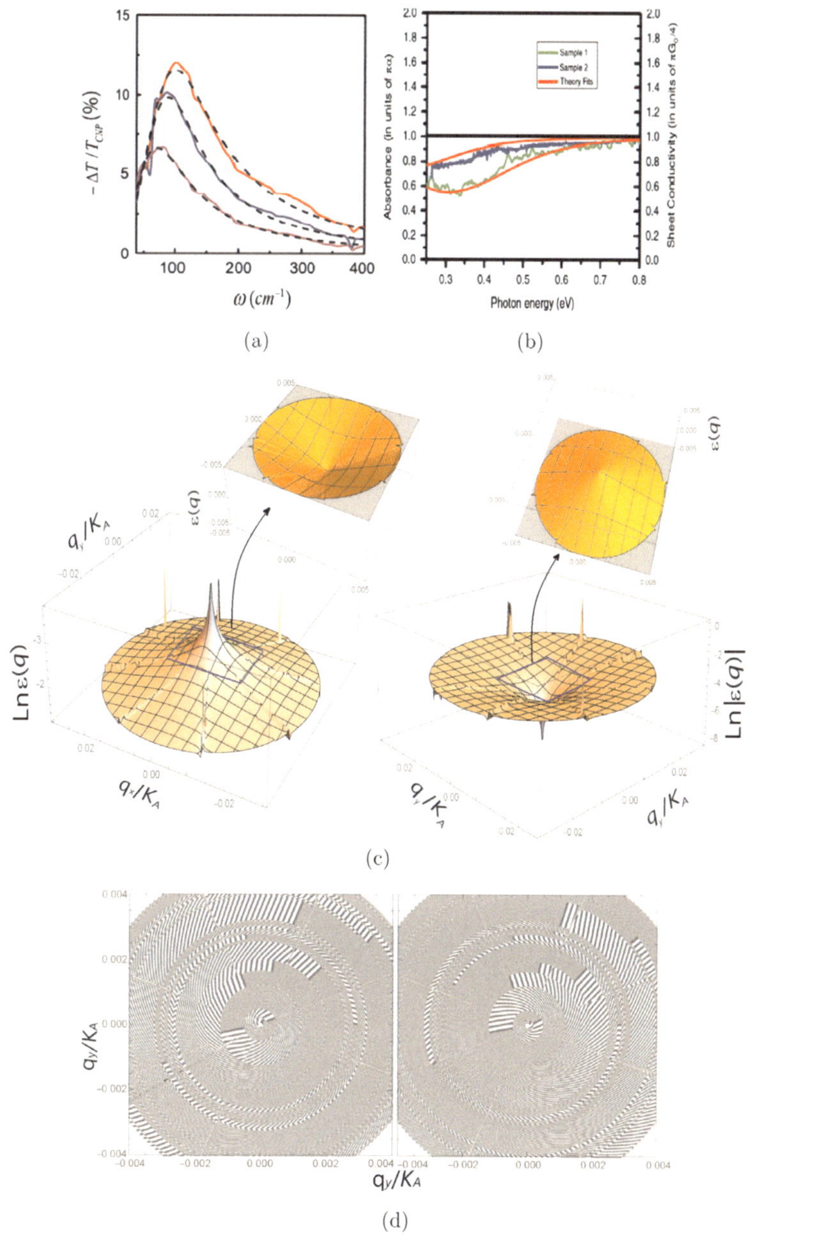


**Figure 1.**
*Graphene conductivity and transmission spectra [10, 12] in the far-infrared (a) and mid-to-near-infrared (b) spectral ranges. Graphene conductivity is obtained from transmittance contrast. The normalized change $\Delta T/T_{CNP}$ in transmittance $T_{CNP}$ of the non-doped graphene is induced by the presence of the graphene sample; $T_{CNP}$ is the transmittance at the applied voltage for charge-neutral Dirac point. The gate-induced change of transmission in graphene is obtained for different values of gate voltage. The Drude conductivity fitting is shown. (c) Electron (left) and hole bands (right) of a quasi-relativistic $N_F = 3$-flavors model graphene that were calculated with a pseudo-Majorana mass term. (d) The vortex texture in contour plots of electron (left) and hole (right) bands was calculated with the Majorana-like mass term on momentum scales $q/K_A = 0.002$; $\vec{q} = \vec{k} - \vec{K}$.*

to perform simulations of the complex conductivity of the system in a number of experimentally interesting cases. The Wilson non-closed loop method will be used to prove dichroism of the band structure, which leads to valley Hall conductivity.

## 2. Fundamentals

Quasi-relativistic graphene model has been derived in [19] as a consequent account of the effect of relativistic exchange interactions, while being grounded on truly secondary quantized relativistic consideration of the problem within the known Dirac-Hartree-Fock self-consistent field approximation. In subsequent publications [23, 27, 28], it has been established that the model admits a form as a Majorana-like system of equations as well as a two-dimensional Dirac-like equation with an additional "Majorana-force correction" term [23]. It reads

$$v_F\left[\vec{\sigma}_{2D}^{AB}\cdot\vec{p}_{BA}-c^{-1}M_{BA}\right]\left|\psi_{BA}^{*}\right\rangle=E\left|\psi_{BA}^{*}\right\rangle \tag{1}$$

and the same equation with labels $(AB, BA)$ exchanged for another sublattice. Here, $(AB, BA)$ are related to sublattices and refer to the quantities that are obtained by similar transformations with a relativistic exchange matrix $\Sigma_{rel}^{x}$; for example, for the momentum operator $\vec{p}$ one gets $\vec{p}_{BA}=\Sigma_{BA}\ \vec{p}\ \Sigma_{BA}^{-1}$; $v_F$ is the Fermi velocity. The vector of two-dimensional (2D) Pauli matrixes comprises two matrixes $\sigma_{2D}=\left(\sigma_x,\sigma_y\right)$. The term $M_{BA}=-\frac{1}{cv_F}\Sigma_{BA}\Sigma_{AB}$ is a Majorana-like mass term, where $c$ is the speed of light. It turns out to be zero in the Dirac point $K(K')$ and gives a very small momentum-dependent correction outside of $K(K')$. The relativistic exchange operator for tight-binding approximation and accounting of nearest lattice neighbors is given by its action on secondary quantized wave functions on sublattices $A(B)$ of the system [19, 27, 28]

$$\begin{aligned}\Sigma_{rel}^{x}&\begin{pmatrix}\hat{\chi}_{-\sigma A}{}^{\dagger}\left(\vec{r}\right)\\ \hat{\chi}_{\sigma_B}^{\dagger}\left(\vec{r}\right)\end{pmatrix}|0,-\sigma\rangle|0,\sigma\rangle\\ &=\begin{pmatrix}0&\Sigma_{AB}\\ \Sigma_{BA}&0\end{pmatrix}\begin{pmatrix}\hat{\chi}_{-\sigma_A}^{\dagger}\left(\vec{r}\right)\\ \hat{\chi}_{\sigma_B}^{\dagger}\left(\vec{r}\right)\end{pmatrix}|0,-\sigma\rangle|0,\sigma\rangle,\end{aligned} \tag{2}$$

$$\begin{aligned}\Sigma_{AB}\hat{\chi}_{\sigma_B}^{\dagger}\left(\vec{r}\right)|0,\sigma\rangle&=\sum_{i=1}^{N_v N}\int d\vec{r}_i\hat{\chi}_{\sigma_i^B}^{\dagger}\left(\vec{r}\right)|0,\sigma\rangle\Delta_{AB}\\ &\left\langle 0,-\sigma_i\left|\hat{\chi}_{-\sigma_i^A}^{\dagger}\left(\vec{r}_i\right)V\left(\vec{r}_i-\vec{r}\right)\hat{\chi}_{-\sigma_B}\left(\vec{r}_i\right)\right|0,-\sigma_{i'}\right\rangle,\end{aligned} \tag{3}$$

$$\begin{aligned}\Sigma_{BA}\hat{\chi}_{-\sigma_A\dagger}\left(\vec{r}\right)|0,-\sigma\rangle&=\sum_{i'=1}^{N_v N}\int d\vec{r}_{i'}\hat{\chi}_{-\sigma_{i'}^A\dagger}\left(\vec{r}\right)|0,-\sigma\rangle\Delta_{BA}\\ &\left\langle 0,\sigma_{i'}\left|\hat{\chi}_{\sigma_{i'}^B}^{\dagger}\left(\vec{r}_{i'}\right)V\left(\vec{r}_{i'}-\vec{r}\right)\hat{\chi}_{\sigma_A}\left(\vec{r}_{i'}\right)\right|0,\sigma_i\right\rangle.\end{aligned} \tag{4}$$

Here, interaction $(2\times 2)$-matrices $\Delta_{AB}$ and $\Delta_{BA}$ are gauge fields (or components of a gauge field). Vector-potentials for these gauge fields are introduced by the phases $\alpha_0$

and $\alpha_{\pm,k}$, $k = 1, \; 2, \; 3$ of $\pi(p_z)$-electron wave functions $\psi_{\;p_z}\left(\vec{r}\right)$ and $\psi_{p_z, \pm\vec{\delta}_k}\left(\vec{r}\right)$ attributed to a given lattice site and its three nearest neighbors (see details in [28]), $V\left(\vec{r}\right)$ is the three-dimensional (3D) Coulomb potential, and the summation is performed on all lattice sites and number of electrons. The introduction of these three non-Abelian gauge fields was stipulated by a requirement of the reality of eigenvalues of the Hamiltonian operator as gauge conditions. In this case, the operator of relativistic exchange gains an additional implicit $\vec{k}$-dependence upon momentum in the case of nonzero values of gauge fields.

## 3. Electronic structure

The band structure of graphene within the quasi-relativistic $N_F = 3$-flavors model has been calculated (see [25] and references therein) with the Majorana-like mass term and is presented in **Figure 1c**. The graphene bands are conical near the Dirac point at $q(q') \to 0$, $q = \left|\vec{p} - \vec{K}\right|$ $\left(q' = \left|\vec{p}' - \vec{K}'\right|\right)$ where $\vec{p}\left(\vec{p}'\right)$ is a momentum of electron (hole). But, they flatten at large $q(q')$. The band structure of graphene within the quasi-relativistic model with pseudo-Majorana charge carriers hosts vortex and antivortex whose cores are in the graphene valleys $\vec{K}$ and $\vec{K}'$ of the Brillouin zone, respectively (**Figure 1c** and **d**). Touching the Dirac point $K(K')$, the cone-shaped valence and conduction bands of graphene are flattened at large momenta $p$ of the graphene charge carriers [23]. It signifies that the Fermi velocity $v_F$ diminishes drastically to very small values at large momenta $p$. Since eight sub-replicas of the graphene band near the Dirac point degenerate into the eightfold conic band

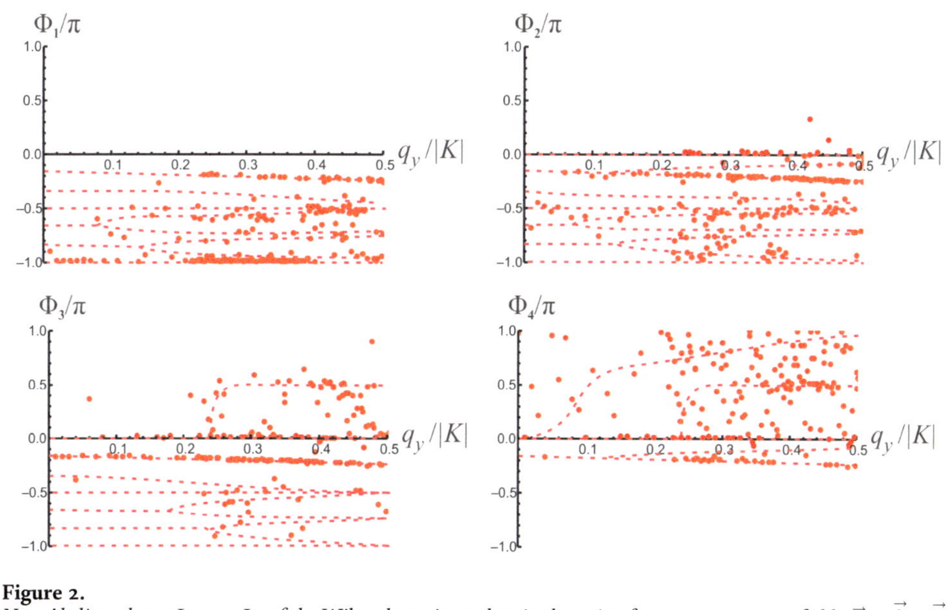


**Figure 2.**
*Non-Abelian phases $\Phi_1, \ldots, \Phi_4$ of the Wilson loop eigenvalues in the units of $\pi$ at nonzero gauge fields; $\vec{q} = \vec{k} - \vec{K}$.*

(**Figure 1c** and **d**) the pseudo-Majorana fermions forming the eightfold degenerate vortex are confined by the hexagonal symmetry. In the state of confinement, the pseudo-Majorana fermions are linked with the formation of electron-hole pairs under the action of the hexagonal symmetry.

A global characterization of all Dirac touching with non-Abelian Zak phases $\Phi_1, \Phi_2, \Phi_3, \Phi_4$ as arguments of the Wilson loop operator for our model is presented in **Figure 2** [23]. Accordingly, in the case of the nonzero gauge fields, the simulations predict the homotopy group $Z_{12}$ with the generator $-\pi/6$ in the Dirac points $K(K')$ [23]. **Figure 1d** shows that the proposed model has the nontrivial topological properties revealing in the dichroism of its band structure. The vorticity of the band is originated from Majorana-like excitations. It is natural to assume that such peculiarities would lead also to observable consequences in charge transport in such a system.

## 4. Non-Abelian currents in quasi-relativistic graphene model

Conductivity can be considered as a coefficient linking the current density with an applied electric field in a linear regime of response. To reach the goal, several steps should be performed. First, one has to subject the system to an electromagnetic field, and this can be implemented by a standard change to canonical momentum $\vec{p} \to \vec{p} - \frac{e}{c}\vec{A}$ in the Hamiltonian in the following way:

$$\left[c\vec{\sigma}_{2D}^{BA} \cdot \left(\vec{p}_{AB} - \frac{e}{c}\vec{A}\right) - \widetilde{\Sigma_{AB}\Sigma_{BA}}\left(\vec{p}_{AB} - e\vec{A}/c\right)\right]\widehat{\widetilde{\chi}^{\dagger}_{+\sigma_B}}\left(\vec{r}\right)\left|0{-}\sigma\right\rangle = cE_{qu}(p)\widehat{\widetilde{\chi}^{\dagger}_{+\sigma_B}}\left(\vec{r}\right)\left|0, -\sigma\right\rangle,$$

$$\widetilde{\Sigma_{AB}\Sigma_{BA}}\left(\vec{p}_{AB} - e\vec{A}/c\right) = \widetilde{\Sigma_{AB}\Sigma_{BA}}(0) + \sum_i \frac{d\widetilde{\Sigma_{AB}\Sigma_{BA}}}{dp'_i}\Big|_{p'_i=0}\left(p_i^{AB} - \frac{e}{c}A_i\right) \tag{5}$$

$$+\frac{1}{2}\sum_{i,j}\frac{d^2\widetilde{\Sigma_{AB}\Sigma_{BA}}}{dp'_i dp'_j}\Bigg|_{p'_i,\ p'_j=0}\left(p_i^{AB} - \frac{e}{c}A_i\right)\left(p_j^{AB} - \frac{e}{c}A_j\right) + \dots . \tag{6}$$

Here, $\vec{A}$ is a vector potential of the field. In what follows, we omit the cumbersome designation "AB" if this does not lead to a lack of sense.

Second, to find the quasi-particle current we use the perturbation theory [29, 30]. Taking into account the expression (6), a potential energy operator $V$ for interaction between the secondary quantized fermionic field $\chi_{+\sigma_B}(x)$ with the electromagnetic field reads

$$V = \begin{array}{l} \chi^{\dagger}_{+\sigma_B}\left[-c\vec{\sigma}_{BA}\cdot\frac{e}{c}\vec{A} - M_{BA}(0) - \sum_i \frac{dM_{BA}}{dp'_i}\Big|_{p'_i=0} \times \left(p_i^{AB} - \frac{e}{c}A_i\right)\right. \\ \left. - 12\sum_{i,j}\frac{d^2M_{BA}}{dp'_i dp'_j}\Big|_{p'_i,\ p'_j=0} \times \left(p_i^{AB} - \frac{e}{c}A_i\right)\left(p_j^{AB} - \frac{e}{c}A_j\right) + \dots\right]\chi_{+\sigma_B}. \end{array} \tag{7}$$

Then, one can find a quasi-relativistic current [31] of charge carriers in graphene as:

$$
\begin{aligned}
j_i^{SM} &= c^{-1} j_i \equiv j_i^{O} + j_i^{Zb} + j_i^{so}, \\
j_i &= e\chi^{\dagger}_{+\sigma_B}(x^+) v^i_{x^+x^-} \chi_{+\sigma_B}(x^-) \\
&- \frac{e^2 A_i}{c M_{AB}} \chi^{\dagger}_{+\sigma_B}(x^+) \chi_{+\sigma_B}(x^-) \\
&+ \frac{e\hbar}{2M_{AB}} \left[ \vec{\nabla} \times \chi^{\dagger}_{+\sigma_B}(x^+) \vec{\sigma} \chi_{+\sigma_B}(x^-) \right]_i, \quad i = 1, 2.
\end{aligned}
\tag{8}
$$

Here,

$$
x^{\pm} = x \pm \varepsilon, \ x = \left\{ \vec{r}, \ t_0 \right\}, \ \vec{r} = \{x, y\}, \ t_0 = 0, \ \varepsilon \to 0;
\tag{9}
$$

$v^i_{x^+x^-}$ is the velocity operator determined by a derivative of the Hamiltonian (1), $\chi_{+\sigma_B}(x^+)$ is the secondary quantized fermion field, the terms $j_i^O, \ j_i^{Zb}, \ j_i^{so}, i = x, y$ describe an ohmic contribution that satisfies the Ohm's law and contributions of the polarization and magneto-electric effects, respectively.

To perform quantum-statistical averaging for the case of nonzero temperature, we use a quantum-field method developed in references [32, 33]. After tedious but simple algebra one can find the conductivity in our model:

$$
\begin{aligned}
&\sigma_{ii}^{O}(\omega, \ k) \\
&= \frac{ie^2 \overline{\beta}^2}{(2\pi c)^2} \mathrm{Tr} \int \left( 1 - M_{BA}\left(\vec{p}\right) \frac{\partial^2 M_{BA}}{\partial p_i^2} \right) \\
&\times \left( M\vec{v}^{\,i}(p), N\vec{v}^{\,i}(p) \right) d\vec{p},
\end{aligned}
\tag{10}
$$

$$
\begin{aligned}
&\sigma_{ll}^{Zb}(\omega, k) = \frac{ie^2}{\overline{\beta}^2 (2\pi c)^2} \\
&\times \mathrm{Tr} \int \frac{M_{BA}\left(\vec{p}\right)}{2} \sum_{i=1}^{2} \frac{\partial^2 M_{BA}}{\partial p_i^2} \left( M\vec{v}^{\,i}(p), N\vec{v}^{\,i}(p) \right) d\vec{p},
\end{aligned}
\tag{11}
$$

$$
\begin{aligned}
&\sigma_{12(21)}^{so}(\omega, k) = (-1)^{1(2)} \frac{i}{2} \frac{ie^2 \overline{\beta}^2}{(2\pi c)^2} \\
&\times \mathrm{Tr} \int M_{BA}\left(\vec{p}\right) \frac{\partial^2 M_{BA}}{\partial p_1 \partial p_2} \left( M\vec{v}^{\,1(2)}(p), N\vec{v}^{\,1(2)}(p) \right) \sigma_z d\vec{p}
\end{aligned}
\tag{12}
$$

for the current $j_i^O, \ j_i^{Zb}, \ j_i^{so}$, respectively. Here, matrices $M, \ N$ are given by the following expressions:

$$
\begin{aligned}
M &= \frac{f\left[\overline{\beta}((H(p^+) - \mu)/\hbar)\right] - f\left[\overline{\beta}(H^{\dagger}(-p^-) - \mu/\hbar)\right]}{\overline{\beta}z - \overline{\beta}(H(p^+)/\hbar) + \overline{\beta}(H^{\dagger}(-p^-)/\hbar)}, \\
N &= \frac{\delta(\hbar\omega + \mu)}{(\hbar z + H(p^+) - H^{\dagger}(-p^-))\overline{\beta}}.
\end{aligned}
$$

Here, $f$ is the Fermi–Dirac distribution, $z = \omega + i\varepsilon$, $\vec{p}^{\pm} = \vec{p} \pm \vec{k}$, $\omega$ is a frequency, $\mu$ is a chemical potential, and $\overline{\beta}$ is an inverse temperature divided by $c$.

## 5. Method of conductivity calculation

### 5.1 Diagonalized representation of the Hamiltonian

We perform the conductivity calculation in a representation where the Hamiltonian of the graphene model is diagonalized to find the velocity determined in the following way:

$$\vec{v}_{AB} \approx \frac{\partial H_{AB}}{\partial \vec{p}}. \tag{13}$$

The velocity operator $\vec{v}_{AB(BA)}$ should be transformed by the transformation of the form $S^{-1}\vec{v}_{AB}S$ with a matrix $S$ constructed on the eigenvectors $\chi$ of the Hamiltonian. The corresponding matrix $S$ should be constructed on the eigenvectors of the operator adjoined to the Hamiltonian. In every $p$-point the particle (hole) Hamiltonian is represented by $2 \times 2$ matrix, we denote matrix elements of the exchange operator $\left(i\Sigma^{x}_{rel}\right)_{AB(BA)}$ formally as $\Sigma^{AB(BA)}_{ij}$. Then, the eigenvectors $\chi_i$, $i = 1, 2$ of the Hamiltonian (1) being the rows of the appropriate matrix $S$ can be expressed in an explicit way:

$$\chi_1 = \left\{ \frac{ip\sin(\phi)\left(\Sigma^{AB}_{11}\Sigma^{AB}_{21} + \Sigma^{AB}_{12}\Sigma^{AB}_{22}\right) + p\cos(\phi)\left(\Sigma^{AB}_{12}\Sigma^{AB}_{22} - \Sigma^{AB}_{11}\Sigma^{AB}_{21}\right) - p\left(\Sigma^{AB}_{12}\Sigma^{AB}_{21} - \Sigma^{AB}_{11}\Sigma^{AB}_{22}\right)}{p\left(\left(\left(\Sigma^{AB}_{11}\right)^2 - \left(\Sigma^{AB}_{12}\right)^2\right)\cos(\phi) - i\left(\left(\Sigma^{AB}_{11}\right)^2 + \left(\Sigma^{AB}_{12}\right)^2\right)\sin(\phi)\right)}, 1 \right\} \tag{14}$$

$$\chi_2 = \left\{ \frac{ip\sin(\phi)\left(\Sigma^{AB}_{11}\Sigma^{AB}_{21} + \Sigma^{AB}_{12}\Sigma^{AB}_{22}\right) + p\cos(\phi)\left(\Sigma^{AB}_{12}\Sigma^{AB}_{22} - \Sigma^{AB}_{11}\Sigma^{AB}_{21}\right) + p\left(\Sigma^{AB}_{12}\Sigma^{AB}_{21} - \Sigma^{AB}_{11}\Sigma^{AB}_{22}\right)}{p\left(\left(\left(\Sigma^{AB}_{11}\right)^2 - \left(\Sigma^{AB}_{12}\right)^2\right)\cos(\phi) - i\left(\left(\Sigma^{AB}_{11}\right)^2 + \left(\Sigma^{AB}_{12}\right)^2\right)\sin(\phi)\right)}, 1 \right\} \tag{15}$$

In this way, we can calculate numerically the velocity operator in every $p$-point with subsequent its substitution to the conductivity integral.

### 5.2 Integral calculations

Contributions to conductivity include 2D integrals over the Brillouin zone (BZ). For example, the integrals in the ohmic contribution given by formulae (10) have the following form:

$$\sigma^{intra}_{ij}\left(\omega, \vec{k}\right) = \sum_{a=1,2} \frac{ie^2 v_0^2}{\pi^2} \int_{BZ} d^2\vec{p} \frac{\left(\left[v^i_{aa}\left(\vec{p}\right)v^j_{aa}\left(\vec{p}\right)\right]f\left[\varepsilon_1\left(\vec{p} - \vec{k}/2\right)\right] - f\left[\varepsilon_1\left(\vec{p} + \vec{k}/2\right)\right]\right)}{\left(\varepsilon_1\left(\vec{p} + \vec{k}/2\right) - \varepsilon_1\left(\vec{p} - \vec{k}/2\right)\right)\left(\omega - \varepsilon_1\left(\vec{p} + \vec{k}/2\right) + \varepsilon_1\left(\vec{p} - \vec{k}/2\right)\right)}, \tag{16}$$

$$\sigma_{ij}^{inter}\left(\omega,\vec{k}\right)=\frac{2i\omega e^2v_0^2}{\pi^2}\int\limits_{BZ}d^2\vec{p}\frac{v_{12}^i\left(\vec{p}\right)v_{21}^i\left(\vec{p}\right)\left(f\left[\varepsilon_1\left(\vec{p}-\vec{k}/2\right)\right]-f\left[\varepsilon_2\left(\vec{p}+\vec{k}/2\right)\right]\right)}{\left(\varepsilon_2\left(\vec{p}+\vec{k}/2\right)-\varepsilon_1\left(\vec{p}-\vec{k}/2\right)\right)\left(\omega^2-\left(\varepsilon_2\left(\vec{p}+\vec{k}/2\right)-\varepsilon_1\left(\vec{p}-\vec{k}/2\right)\right)^2\right)}. \tag{17}$$

Here, the first integral is for the intraband transitions, and the second one is for the interband ones. Now, we highlight the pole structure for the integrands for small but finite $\vec{k}$, accounting for that $\varepsilon_1\left(\vec{p}\right)=-\varepsilon_2\left(\vec{p}\right)$. The first integral can be rewritten as:

$$\sigma_{ij}^{intra}\left(\omega,\vec{k}\right)=\sum_{a=1,2}\frac{ie^2v_0^2}{\pi^2}\int\limits_{BZ}d^2\vec{p}\left[v_{aa}^i\left(\vec{p}\right)v_{aa}^j\left(\vec{p}\right)\right]\frac{df(\varepsilon)}{d\varepsilon}\bigg|_{\varepsilon=\varepsilon_1(\vec{p})}\frac{1}{\left(\omega-\vec{k}\cdot\nabla\varepsilon_1\left(\vec{p}\right)\right)}, \tag{18}$$

where we have performed the Taylor series expansion on $|\vec{k}|$ up to linear terms. In the second integral, only the second term in the denominator can produce poles, so we expand it into a power series on $|\vec{k}|$, making a change to polar coordinates $\left(p_x,p_y\right)\to(p,\phi)$ that results:

$$\sigma_{ij}^{inter}\left(\omega,\vec{k}\right)=\frac{2i\omega e^2v_0^2}{\pi^2}\int\limits_{BZ}pdpd\phi\frac{v_{12}^i\left(\vec{p}\right)v_{21}^i\left(\vec{p}\right)\left(f\left[\varepsilon_1\left(\vec{p}-\vec{k}/2\right)\right]-f\left[\varepsilon_2\left(\vec{p}+\vec{k}/2\right)\right]\right)}{\left(\varepsilon_2\left(\vec{p}+\vec{k}/2\right)-\varepsilon_1\left(\vec{p}-\vec{k}/2\right)\right)\left(\omega^2-4p^2-k^2\sin^2\phi\right)}. \tag{19}$$

Pole structure of (18) and (19) is presented in **Figure 3**. In accordance with Eq. (18) and **Figure 3** at $\vec{k}=0$ this integral is a regular one, whereas at finite $k$ there is a line of poles (dashed lines in the **Figure 3**). At a finite $k$ the pole structure of $\sigma_{ij}^{inter}\left(\omega,\vec{k}\right)$ (19) is an elliptic one that results in the necessity to account for an infinite sum of poles as a contribution to conductivity: $\sigma_{ij}^{inter}\left(\omega,\vec{k}\right)\propto\int d\phi\ \ \mathrm{Re\,s}(\phi)$, where $Res(\phi)$ is a residue in the pole located at angle $\phi$ on the poles line. The integral (19) in the case $k=0$ holds poles laying at a circumference that can be effectively reduced to a single one as $\sigma_{ij}^{inter}\left(\omega,\vec{k}\right)\propto 2\pi\ \ \mathrm{Re\,s_1}$, where $\mathrm{Re\,s_1}$ is a residue in arbitrary point of the circumference. For every oblate ellipse at large wave numbers, $k$, the main contribution to the integral (19) gives the only points touching the circumference $\sigma_{ij}^{inter}\left(\omega,\vec{k}\right)\sim 2\ \ \mathrm{Re\,s_1}$. Thus, the value of the optical conductivity decreases with the growth of $k$.

We define the upper integration limit for the model with linear dispersion in a way based on reasoning from the energy limit of tight-binding-approximation applicability ($\omega_{max}<1$ eV $\approx 10^4$ K). Then, the upper integration limit $q_{max}$ on momentum is $q_{max}\sim\omega_{max}/v_F$, $\omega_{max}<1$ eV, and correspondingly for the massless Dirac fermions model, the integration should use the range from 0 to $|\vec{q}|<0.14\,|\vec{K}_A|$. As the simulation results presented in **Figure 4** show the integration within this limit leads to the conductivity fall in the range $\omega_{max}\sim 4000$ K (0.3 eV) at the temperature of 3 K.

Thus, the range of momenta to predict conductivity in near-infrared and visible spectral ranges is outside the limits of applicability of the massless pseudo-Dirac fermion model.

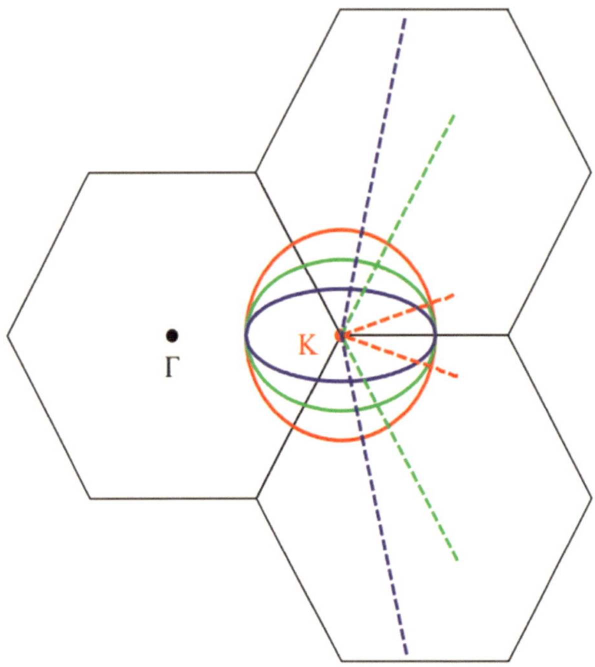


**Figure 3.**
*Sketch of the poles structure of integrands for intra-(dashed lines) and inter-(solid lines) bands contribution to conductivity. Solid lines parameters at $\omega = 1$ (in units of $v_F|\vec{K}_A|$) are $k/|\vec{K}_A| = 0.1$ (red), 0.7 (green), 0.9 (blue). Dashed line parameters at $\omega = 0.2$ (in units of $v_F|\vec{K}_A|$) are $k/|\vec{K}_A| = 0.21$ (red), 0.4 (green), 0.9 (blue).*

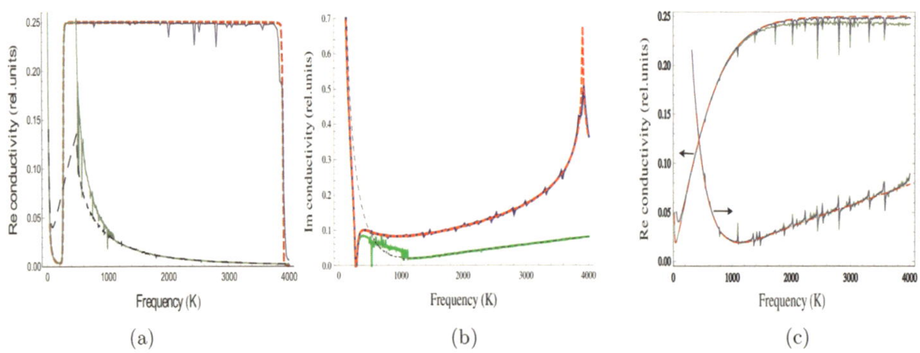


**Figure 4.**
*Frequency dependencies of the real and imaginary parts of the massless ohmic term of conductivity at very small wave number $q = 10^{-8}|\vec{K}_A|$ in relative units of $e^2/\hbar$; a cutting parameter $q_{max} = 0.14\,|\vec{K}_A|$. The model [9, 32] is simulated at $T = 3$ K, chemical potential $\mu = 135K$ (red-dashed lines in figures (a–b)) and at $T = 200$ K, $\mu = 33$ K (red-dashed line in figure (c)). Numerical results for our model are green-solid and black-dashed lines for the approximation with zero gauge phases at $T = 3$ K, $\mu = 135K$ and $T = 200$ K, $\mu = 33K$, respectively, and blue line for the approximation with nonzero gauge phases at $T = 3$ K, $\mu = 135K$ in figures (a–b). (c) Dependence of the real and imaginary parts of the massless ohmic term of conductivity on the damping $\gamma$ for the approximation of the nonzero gauge fields with $\gamma = 0.1$ (green curve) and $\gamma = 1$ (blue curve) at $T = 200$ K, $\mu = 33K$.*

According to our calculation presented in **Figure 1c**, the energy bandstructure is flattened for the pseudo-Majorana fermion graphene model. It signifies that the Fermi velocity $v_M$ for our model pseudo-Majorana states tends to zero for high wave numbers $q$. Since $v_M \to 0$ for large $q$ the applicability condition $v_M q < \omega_{max}$ holds always. We choose $|\vec{q}| \leq 0.38|\vec{K}_A|$ as the limit of integration over momentum $|\vec{q}|$ because it is in the range of $|\vec{q}|$ from $0.28|\vec{K}_A|$ to $0.56|\vec{K}_A|$ appropriate for our model. Such choice corresponds to $\omega_{max} > 7000$ K.

In the approximation with zero-phases of the gauge fields, the analytical formulas for the integrands in the conductivity contribution terms have been used. The integrals have been calculated with adaptive integration steps in both directions $(|k|, \phi)$ providing high calculation accuracy (not less than 0.01%).

In the approximation with nonzero gauge fields, we have to calculate numerically by introducing into consideration the small-positive damping constant $\gamma$ for the states as a small imaginary contribution to the energies. The values of $\gamma$ define the extent of smoothness of the singular behavior of the integrand and does not influence the general form of the dependency curve in accordance with **Figure 4c**.

All quantities necessary for the calculation of the complex conductivity have been calculated on a grid in the space of wave vectors with 200-point discretization in the angle $\phi$ for every given wave number $q$ and variable step (a denser grid at small wave numbers and larger at large ones) to the maximum wave number $q_{max} = 0.38|\vec{K}_A|$. The 2D interpolation on this grid has been used for integrands evaluation in the intermediate points that are necessary for conductivity simulations. An error stipulated by the interpolation from the grid in wave vectors space has been roughly estimated by interpolation of the conic spectrum of the Dirac pseudo-fermion model on the same lattice with subsequent usage of the interpolation data for evaluation of the conductivity. Its value turns out to be less than $10^{-3}$%.

Total estimation of the conductivity error has been performed by variation of the number of points used for interpolation of the energy band spectrum (by diminishing this number at factor two and subsequent comparison of the simulation results in both cases). It turns out to be not exceeding 10% in the considered frequency region. It should be noted that the error bars for values of the Fermi velocity that were measured by different techniques, including transport experiments (Shubnikov–de Haas oscillations) [34], infrared measurements of the Pauli blocking in graphene [35], magneto-optics [36], were also of the order of 10%.

## 6. Far-infrared optical spectroscopy of graphene

In this section, we study optical transitions in the Majorana-like fermion graphene model and compare the theoretical predictions with experimental data for the far-infrared spectral range. The frequency dependencies of the real and imaginary part of the ohmic optical conductivity ($\left|\vec{k}\right| = 0$) for temperatures $T = 3, 200$ K are shown in **Figure 5**. Chemical potentials, $\mu$, are used to be 135 and 33 K for $T = 3$ and 200 K, respectively. For comparison, far-infrared conductivity is shown in the pseudo-Dirac fermion graphene model [9].

The conductivity in the far-infrared spectral range calculated within the massless pseudo-Dirac fermion graphene model gains a constant value that contradicts the experimental data.

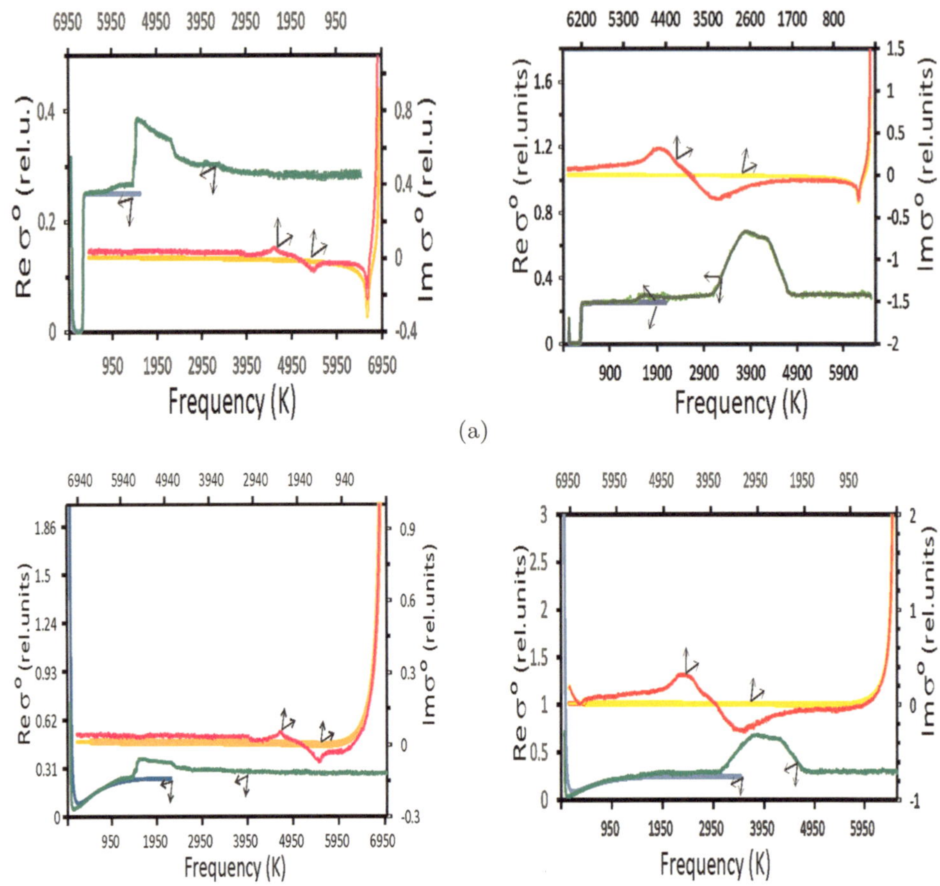


(b)

**Figure 5.**
*Frequency dependencies of real (green and blue solid lines) and imaginary (yellow and magenta solid lines) parts of ohmic contribution $\sigma^{O}_{xx(yy)}$ to the far-infrared conductivity in relative units of $e^2/\hbar$ for our model (1) (green and magenta lines) and for the Dirac massless fermion model [9, 32] (blue and yellow lines). The calculations were carried out at temperatures $T = 3$ (a), 200(b) K and chemical potentials $\mu = 135$ (a), 33(b) K for two case: Taking into account (left) and neglecting (right) the pseudo-Majorana mass term.*

The real part of the ohmic contribution in the optical conductivity for the pseudo-Majorana fermion graphene model undergoes a steep increase finalizing a much slower fall in the spectral range of 1000 – 2000 K (see **Figure 5**, left). The maximum conductivity is at the frequency of ~ 2000 K (0.13 eV). The predicted presence of the asymmetric spectral band in the optical conductivity for graphene is confirmed experimentally that the asymmetric peak is really observed on a frequency of ~ 0.15 eV (~ 150 – 200 $cm^{-1}$) [10].

The slow decrease of the peak intensity compared to the rise finalizes with steep achieving the plateau, which in turn ends with a shallow minimum in the remaining higher-frequency part of the far-infrared spectrum shown in **Figure 5**, left. Verifying the prediction of low (antipeak) being on the frequency of ~ 0.27 eV (3000 K) the experiment [12] confirms that the experimental antipeak is really indicated in the spectral range of ~ 0.2 – 0.3 eV.

The appearance of the peak is accompanied by the tendency of the imaginary part of the ohmic contribution in the optical conductivity toward negative values. It testifies that the position of the peak in the real part correlates with the emergence of plasmon oscillations. The exclusion of the pseudo-Majorana mass term results in the loss of the asymmetry of the far-infrared conductivity peak and a shift to the high-frequency region (see **Figure 5**, right). It testifies that chiral symmetry is restored, and correspondingly the anomalous transport disappears. According to results presented in **Figure 5**, the far-infrared peak and antipeak exist at both low and high temperatures. The influence of temperature consists in their smoothing, which impedes their experimental detection.

## 7. Electronic transport and field effect

In this section, we study essential features of the electric charge transport by Majorana-like carriers in graphene and compare theoretical predictions with experimental data.

Total current $\vec{J}$ in graphene is determined by electron and hole currents of valleys $K, K'$ as $\vec{J} = \vec{J}_K - \vec{J}_{K'}$. The two currents flow on non-coinciding paths in the topologically nontrivial pseudo-Magorana graphene model because the jump of an electron (hole) from a site of sublattice $A$ $(B)$ to the nearest site of sublattice $B$ $(A)$ is equivalent to bypassing the lattice site with the acquisition of the carrier wave function of the phase, being a multiple of the group generator $\pi 6$ of the homotopy group $\mathbf{Z}_{12}$ (**Figure 2**) in addition to 60-degree rotation by virtue of the homotopy group of graphene Brillouin zone and the hexagonal symmetry. It signifies that at transition from one trigonal sublattice to another one, the direction of motion of the charge carrier is rotated at an angle of $\frac{\pi}{2}$. As a result, the $\vec{J}_K, \vec{J}_{K'}$ are orthogonal.

Let us denote the first and the second terms in the conductivity $\sigma_{ii}^{O}$ (see Eq. 10) through $\sigma_{ii}^{o}$ and $\sigma_{ii}^{add}$, respectively: $\sigma_{ii}^{O} = \sigma_{ii}^{o} + \sigma_{ii}^{add}$. $\sigma_{ii}^{add}$ and $\sigma_{ii}^{o}$ depend and do not depend on the Majorana-like mass term $M_{AB}$, respectively. Then, taking into account of the polarization effects (Eq. 11) one can determine the pseudo-Majorana corrections, $\sigma_{ii}^{tp}$, $i = x, y$ to the conductivity in the following way: $\sigma_{ii}^{tp} = \sigma_{ii}^{add} + \sigma_{ii}^{Zb}$, $\sigma_{xx}^{tp} = -\sigma_{yy}^{tp}$.

Then, in the absence of the Majorana conductivity corrections, we get the Ohm's law because of the direction of the sum, $\vec{J}^{o} = \vec{J}_K^{o} - \vec{J}_{K'}^{o}$ of the currents, $\vec{J}_K^{o}, \vec{J}_{K'}^{o}$, equal to $\Re e\ \sigma_{ii}^{o}\vec{E}_i, i = x, y$ coincides with the direction of an applied electric field $\vec{E}$, $\vec{E} = \vec{E}_x + \vec{E}_y$ as **Figure 6** shows. Meanwhile, the direction which is orthogonal to $\vec{E}$ the current is absent. Taking into account the pseudo-Majorana conductivity corrections, the total current changes in the following way: $\vec{J}_{skew} \equiv \vec{J}_K - \vec{J}_{K'} = \vec{J}^{o} + \sigma_{xx}^{tp}\vec{E}_x - \sigma_{yy}^{tp}\vec{E}_y \equiv \sum_{i=1}^{2}(-1)^{i-1}(\sigma_{ii}^{O} + \sigma_{ii}^{Zb})\vec{E}_i$. As **Figure 6** shows the $\vec{J}_{skew}$ is rotated in respect to $\vec{E}$, and correspondingly a nonzero component of the current appears in the direction orthogonal to $\vec{E}$. It proves that a topological current can exist in the pseudo-Majorana graphene model and the abnormal transport appears due to the presence of the Majorana mass term.

**Figure 7** demonstrates negative differential conductivity for the topological current $J^{tp} = \Re e\sigma_{ii}^{tp}(\omega)U$ assuming the increase of the system energy in a form $\hbar\omega \sim U^2$, where $U$ is a bias voltage.

A total spin-orbital valley current $\vec{J}_{VHE} = \sigma^{so}_{xy}\vec{B}_y - \sigma^{so}_{yx}\vec{B}_x$ is produced under an action of a magnetic field $\vec{B}_{\parallel}$ parallel to $\vec{E}$. One can note that $\vec{J}_{VHE}$ is always directed tangentially to the bias $\vec{E}$ and added to the current $\vec{J}^{tp}$. Then, the total current in the

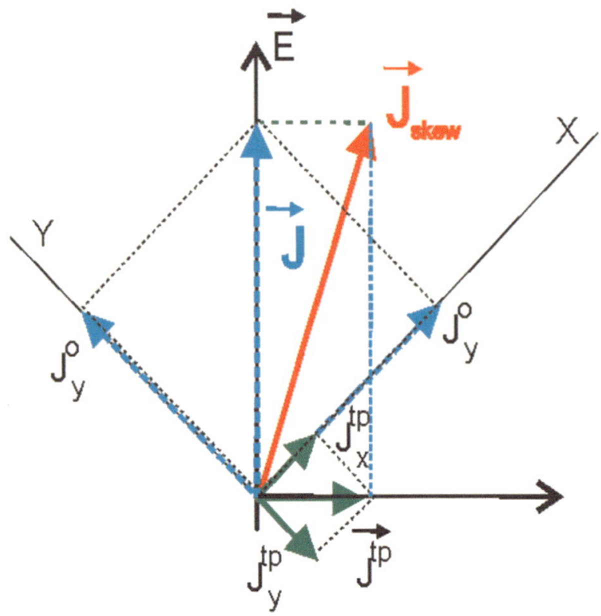


**Figure 6.**
*Sketch of currents in the topologically nontrivial semimetal:* $J^o_{x(y)}$ *is a massless ohmic current along the axis* $X(Y)$, $J^{tp}$ *is a sum,* $\sum_{i=1}^{2}(\sigma^{add}_{ii} + \sigma^{Zb}_{ii})\vec{E}_i$ *of the polarization and dynamical-mass corrections to* $J^o_{x(y)}$. $\vec{E}$ *is an applied electric field.*

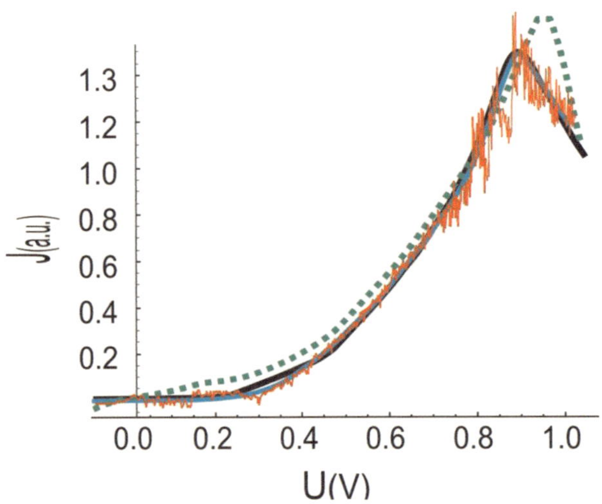


**Figure 7.**
*Dependencies of topological current J on bias voltage U: Simulation results (red solid curve) and their fitting (blue solid curve) for negative differential conductivity (NDC) in our quasi-relativistic graphene model at temperature T = 3 K, chemical potential* $\mu = 135$ *K; black solid and green dashed curves present experimental data and theoretical calculation for NDC in two graphene flakes twisted approximately at 90° to each other at 1.8° misalignment angle [37]. The bias voltage U is given in volts "V."*

direction of the vector $\vec{E}$ increases. It signifies that a negative magnetoresistance (NMR) appears at weak magnetic fields parallel to electric ones. NMR is a specific feature of topological materials and presents a phenomenon of chiral anomaly [38, 39].

Let us investigate longitudinal conductivity for low frequencies $\omega \to 0$ and non-vanishing wave vectors $\vec{k} = \vec{p}/\hbar - \vec{K}_{A(B)}$. The longitudinal conductivity $\sigma_L\left(\omega, \vec{k}\right)$ is determined through the conductivity tensor splitting into longitudinal and transversal terms as [40]

$$\sigma_{ij}\left(\omega, \vec{k}\right) = \left(\delta_{ij} - \frac{k_i k_j}{\vec{k}^2}\right)\sigma_T\left(\omega, \vec{k}\right) + \frac{k_i k_j}{\vec{k}^2}\sigma_L\left(\omega, \vec{k}\right); i,j = x,y. \tag{20}$$

when choosing $\vec{k} = (k_x, 0)$, $\vec{k} = (0, k_y)$, or $k_x = k_y = k$ one always has:

$$\sigma_{xx}\left(\omega, \vec{k}\right) = \sigma_L(\omega, k), \quad or\, \sigma_{yy}\left(\omega, \vec{k}\right) = \sigma_L(\omega, k).$$

Now, let us calculate the low-frequency conductivity $\sigma\left(\omega, \vec{r}\right)$. To do it, one has to perform the inverse Fourier transformation.

$$\sigma\left(\omega, \vec{r}\right) = \frac{1}{(2\pi)^2}\int \sigma_L^o(\omega, k) e^{i\vec{k}\cdot\vec{r}} d^2k. \tag{21}$$

Here, $\omega$ is a cyclic frequency. We consider the effects of spatial dispersion on the real part $\Re e\ \sigma_L^O(\omega, k)$ of the longitudinal complex conductivity at the low frequencies: $\omega = 10^{-10}$, 0.004, 13.3 K (kelvin) for the massless pseudo-Dirac graphene fermion model with the number of flavors $N_F = 2$ (pseudospin and spirality) and our graphene model with the $N_F = 3$ flavors. Conductivity for frequencies in the Hertz range, for example, 2.08 Hz ($\omega = 10^{-10}$ K), can be considered as a minimum conductance for direct current. The numerical results are presented in **Figure 8**.

The function $\Re e\ \sigma_L^o(\omega, k)$ for the massless pseudo-Dirac fermion graphene model is a positive constant function. The $\Re e \sigma_L^o(k)\big|_{k\to\infty} \equiv \Re e\ \sigma_{k_\infty}$ is the positive constant ($\Re e\ \sigma_{k_\infty} > 0$) for the all frequencies (it is equal to $\sim 0.004$ for $\omega = 0.004, 10^{-10}$ K and $\sim 0.017$ for $\omega = 13.3$ K. The real part of the longitudinal complex conductivity in the $N_F = 2$-model becomes the very large positive constant, $\Re e\ \sigma_L^o(k)\big|_{k\to 0} \equiv \Re e\ \sigma_{k_0} \sim$ 44.443, at small values of $k$, $k \ll 1$ for the frequencies $\omega = 10^{-10}$, 0.004 K. As **Figure 8** shows the function $\Re e\ \sigma_L^o(\omega, k)$ in the $N_F = 2$-model sharply increases in a very narrow range of $k/K_A$. Since $\Re e\ \sigma_L^o(\omega, k)$ is constant almost everywhere for the $N_F =$ 2-model of graphene does not oscillate and $e^{i\vec{k}\cdot\vec{r}}$ enters the integrand of the expression (21), the $\Re e\ \sigma\left(\omega, \vec{r}\right)$ is equal to zero, and correspondingly the minimum conductivity is equal to zero in this graphene model. But this prediction for the DC case contradicts the experimental facts that the value of the low-frequency conductivity is in the range $4\frac{e^2}{h}$–$6\frac{e^2}{h}$.

The function $\Re e\ \sigma_L^o(\omega, k)$ for the pseudo-Majorana graphene $N_F = 3$-models both with and without the pseudo-Majorana mass term is a sign-alternating one at $k \to \infty$. It signifies that the plasmon oscillations can emerge in the graphene $N_F = 3$-models.

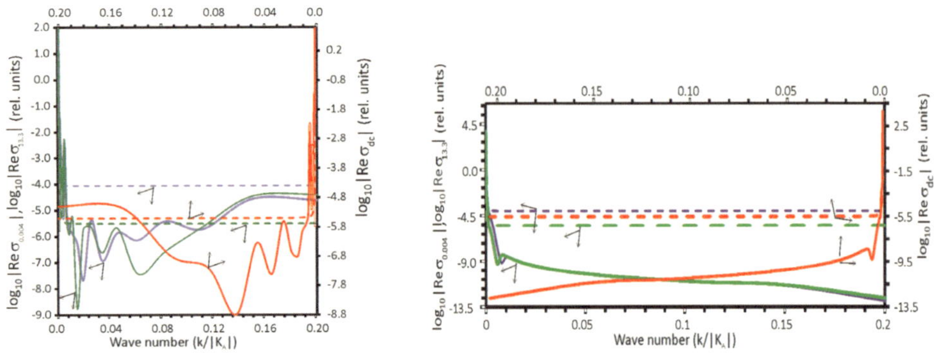


**Figure 8.**
*Real part $\sigma_L^o(\omega, k)$ of the longitudinal ohmic contribution to conductivity vs. wave number k, $\vec{k} = \vec{p} - \vec{K}_{A(B)}$ for our pseudo-Majorana fermion $N_F = 3$-model (1) (solid curves) and for the massless Dirac fermion model [32] (dashed curves), at the frequencies ω: 13.3 K (0.27 THz, blue color), 0.004 K (83 MHz, green color), $10^{-10}$ K (2.08 Hz, red color). The calculations were carried out at temperatures $T = 100$ K and chemical potentials $\mu = 1$ K for two case: Taking into account (left) and neglecting (right) the pseudo-Majorana mass term. $\sigma_L^o(\omega, k)$ is measured in the relative units of $e^2/h$ and labeled as $\Re e\ \sigma_{13.3}$, $\Re e\ \sigma_{0.004}$ and $\Re e\ \sigma_{dc}$ for the frequencies 13.3, 0.004, and $10^{-10}$ K, respectively.*

Neglecting the pseudo-Majorana mass term the function $\Re e\ \sigma_L^o(\omega, k)$ for the pseudo-Majorana fermion graphene $N_F = 3$-model is weakly oscillating for all frequencies (see **Figure 8**). Since $\Re e\ \sigma_L^o(\omega, k)$ for this case is practically constant except for a very narrow interval, then as well as for the $N_F = 2$ model, the $\Re e\ \sigma\left(\omega, \vec{r}\right)$ is equal to zero, and correspondingly the graphene model with the chiral pseudo-Majorana charge carriers is not semimetal.

Now let us examine the $N_F = 3$ model with the nonzero pseudo-Majorana mass term. In this case, the $\Re e\ \sigma_L^o(\omega, k)$ strongly oscillates for all frequencies (**Figure 8**). In this case for $\omega = 0.004, 10^{-10}$ K, since the $\Re e\ \sigma_L^o(\omega, k)$ has the large maximum ($\sim 44.55\frac{e^2}{h}$), and oscillating tends to zero it behaves like the function $\Re e\ \sigma_{k_{max}} \frac{\sin(k - k_{max})}{(k - k_{max})}$. Such sort of functions can be considered as a finite approximation of the Dirac $\delta$−function in the form of the sinc function $\lim_{\varepsilon \to 0^+} \frac{\sin x}{\varepsilon x}$, and the coefficient $\Re e\ \sigma_{k_{max}}$ is called the intensity or spectral power of the $\delta$−function. In the DC limit ($\omega = 10^{-10}$ K), the $\Re e\ \sigma_L^o(\omega, k)$ possesses one maximum at $k \to 0$. Correspondingly, the DC minimum conductivity $\sigma^{dc}\left(\vec{r}\right)$ for the $N_F = 3$ model with the chiral anomaly may be approximated by the integral with only one Dirac $\delta$-function entering the integrand:

$$\sigma^{dc}\left(\omega, \vec{r}\right) = \frac{1}{(2\pi)^2} \int \Re e \sigma_L^o(\omega, k) e^{i\vec{k}\cdot\vec{r}} d^2k \approx \frac{1}{2\pi} \int \Re e \sigma_L^o(\omega, 0) \delta(k) e^{i\vec{k}\cdot\vec{r}} dk = \frac{7e^2}{h}, \quad \omega = 10^{-10}\ K. \tag{22}$$

The minimum DC-conductivity of graphene in devices with a large area of a graphene monolayer on $SiO_2$ turns out to be of $4e^2/h$ [4] at low temperatures ($\sim 1.5$ K). The minimum DC-conductivity of suspended graphene [5, 6] and of graphene on boron nitride substrate [41] is of $\sim 6e^2/h$ at $T \sim 300$ K. Since for $\omega = 0.004$ K one obtains the same value of the conductivity the low-frequency conductivity behaves in

*DOI: http://dx.doi.org/10.5772/intechopen.106144*

a universal manner. Thus, our estimate (22) is in perfect agreement with the experimental data.

## 8. Conclusion

To finalize our finding, within the earlier developed quasi-relativistic model of graphene, the frequency-dependent conductivity of graphene has been investigated theoretically in a number of experimentally interesting cases. It has been demonstrated that accounting of exchange interactions in the model which leads to its topological nontriviality both in electron structure and quasi-particle excitations not only modifies Ohmic conductivity dependencies both on frequency and temperature but also adds few contributions to conductivity stipulated by Majorana-like mass term corrections. Account of these corrections allows one to obtain a fine description of experimentally observable effects in graphene conductivity dependencies as in frequency domain as for knows unsolved problem of minimum DC-conductivity.

## Author details

Halina V. Grushevskaya* and George Krylov
Physics Department, Belarusian State University, Minsk, Belarus

*Address all correspondence to: grushevskaja@bsu.by

## References

[1] Cooper DR, D'Anjou B, Ghattamaneni N, Harack B, Hilke M, Horth A, et al. Experimental review of graphene. ISRN Condensed Matter Physics. **2012;2012**:501686

[2] Pippard AB. Magnetoresistance in Metals. Cambridge: Cambridge University Press; 2009

[3] Hayes IM, Maksimovic N, Lopez GN, Chan MK, Ramshaw BJ, McDonald RD, et al. Superconductivity and quantum criticality linked by the hall effect in a strange metal. Nature Physics. 2021;**17**:58

[4] Novoselov KS, Geim AK, Morozov SV, et al. Electric field in atomically thin carbon films. Science. 2004;**306**:666

[5] Bolotin KI, Sikes KJ, Hone J, Stormer HL, Kim P. Temperature-dependent transport in suspended graphene. Physical Review Letters. 2008; **101**:096802

[6] Du X, Skachko I, Barker A, Andrei EY. Approaching ballistic transport in suspended graphene. Nature Nanotechnology. 2008;**3**:491

[7] Ziegler K. Minimal conductivity of graphene: Nonuniversal values from the Kubo formule. Physical Review B. 2007; **75**:233407

[8] Ando T, Zheng Y, Suzuura H. Dynamical conductivity and zero-mode anomaly in honeycomb lattices. Journal of the Physical Society of Japan. 2002;**71**: 1318-1324

[9] Falkovsky LA. Optical properties of graphene and IV–VI semiconductors. Physics–Uspekhi. 2008; **51**(9):887

[10] Ju L, Geng BS, Horng J, Girit C, Martin M, Hao Z, et al. Graphene plasmonics for tunable terahertz metamaterials. Nature Nanotechnology. 2011;**6**:630-634

[11] Mak KF, Shan J, Heinz TF. Seeing many-body effects in single- and few-layer graphene: Observation of two-dimensional saddle-point excitons. Physical Review Letters. 2011;**106**:046401

[12] Mak KF, Ju L, Wang F, Heinz TF. Optical spectroscopy of graphene: From the far infrared to the ultraviolet. Solid State Communications. 2012;**152**:1341

[13] Mak KF, Sfeir MY, Wu Y, Lui CH, Misewich JA, Heinz TF. Measurement of the optical conductivity of graphene. Physical Review Letters. 2008;**101**:196405

[14] San-Jose P, Lado JL, Aguado R, Guinea F, Fernández-Rossier J. Majorana zero modes in graphene. Physical Review X. 2015;**5**:041042

[15] Cao Y, Fatemi V, Fang S, Watanabe K, Taniguchi T, Kaxiras E, et al. Unconventional superconductivity in magic-angle graphene superlattices. Nature. 2018;**556**:43

[16] Gorbachev RV, Song JCW, Yu GL, Kretinin AV, Withers F, Cao Y, et al. Detecting topological currents in graphene superlattices. Science. 2014; **346**:448

[17] Elias DC, Gorbachev RV, Mayorov AS, Morozov SV, Zhukov AA, Blake P, et al. Dirac cones reshaped by interaction effects in suspended graphene. Nature Physics. 2012;**8**:172

[18] Wang J-R, Liu G-Z. Eliashberg theory of excitonic insulating transition

in graphene. Journal of Physics. 2011;**23**: 155602

[19] Grushevskaya HV, Krylov GG. Electronic structure and transport in graphene: QuasiRelativistic Dirac-Hartree-Fock self-consistent field approximation. In: Aliofkhazraei M et al., editors. Graphene Science Handbook: Electrical and Optical Properties. Vol. 3. USA, UK: Taylor and Francis Group, CRC Press; 2016

[20] Grushevskaya H, Krylov G. Massless Majorana-like charged carriers in two-dimensional semimetals. Symmetry. 2016;**8**:60

[21] Grushevskaya HV, Krylov GG. Low frequency conductivity in monolayer graphene model with partial unfolding of Dirac bands. International Journal of Modeling Physics. 2016;**30**:1642009

[22] Grushevskaya HV, Krylov GG. Increasing spin-orbital coupling at relativistic exchange interaction of electron-hole pairs in graphene. Semiconductors. 2018;**52**:1879

[23] Grushevskaya H, Krylov G. Vortex dynamics of charge carriers in the quasi-relativistic graphene model: High-energy $\vec{k} \cdot \vec{p}$ approximation. Symmetry. 2020; **12**:261

[24] Grushevskaya HV, Krylov GG. Non-abelian properties of charge carriers in a quasirelativistic graphene model. Semiconductors. 2020;**54**:1737-1739

[25] Grushevskaya H, Krylov G. Polarization in quasirelativistic graphene model with topologically non-trivial charge carriers. Quantum Reports. 2022;**4**:1

[26] Grushevskaya HV, Krylov GG, Kruchinin SP, Vlahovic B, Bellucci S. Electronic properties and quasi-zero-energy states of graphene quantum dots. Physical Review B. 2021;**103**:235102

[27] Grushevskaya HV, Krylov G. Semimetals with Fermi velocity affected by exchange interactions: Two dimensional Majorana charge carriers. International Journal of Nonlinear Phenomenon in Complex System. 2015; **18**:266

[28] Grushevskaya HV, Krylov G, Gaisyonok VA, Serow DV. Symmetry of model N = 3 for graphene with charged Pseudo-excitons. International Journal of Nonlinear Phenomenon in Complex System. 2015;**18**:81

[29] Dyson F. Advanced Quantum Mechanics. Singapore: World Scientific Publishing; 2007

[30] Krylova H, Gursky L. Spin Polarization in Strongly Correlated Systems. Saarbrucken, Germany: LAP Lambert Academic Publishing; 2013

[31] Davydov AS. Quantum Mechanics. Moscow: Science; 1973

[32] Falkovsky LA, Varlamov AA. Space-time dispersion of graphene conductivity. European Physics Journal. 2007;**56**:281

[33] Grushevskaya HV, George Krylov VA, Gaisyonok. Non-abelian currents in quasi-relativistic graphene model: General theory. International Journal of Nonlinear Phenomenon in Complex System. 2018;**21**:153

[34] Elias DC, Gorbachev RV, Mayorov AS, Morozov SV, Zhukov AA, Blake P, et al. Dirac cones reshaped by interaction effects in suspended graphene. Nature Physics. 2011;7:701

[35] Li ZQ et al. Nature Physics. 2008;**4**: 532

[36] Chen Z-G, Shi Z, Yang W, Lu X, Lai HY, Wang F, et al. Observation of an intrinsic bandgap and Landau level renormalization in graphene/boron-nitride heterostructures. Nature. Communications. 2014;**5**:4461

[37] Mishchenko A, Tu JS, Cao Y, Gorbachev RV, Wallbank JR, Greenaway MT, et al. Twist-controlled resonant tunnelling in graphene/boron nitride/graphene heterostructures. Nature Nanotechnology. 2014;**9**:808

[38] Lu H-Z, Shen S-Q. Quantum transport in topological semimetals under magnetic fields. Frontiers of Physics. 2017;**12**(3):127201

[39] Niemann AC, Gooth J, Wu S-C, et al. Chiral magnetoresistance in the Weyl semimetal NbP. Scientific Reports. 2017;**7**:43394

[40] Kraeft WD, Kremp D, Ebeling W, Röpke G. Quantum Statistics of Charged Particle Systems. Berlin: Academie-Verlag; 1986

[41] Dean CR, Young AF, Meric I, et al. Boron nitride substrates for high-quality graphene electronics. Nature Nanotechnology. 2010;**5**:722

Chapter 8

# Graphene Exfoliation from HOPG Using the Difference in Binding Energy between Graphite, Graphene and a Substrate

*Yannick A.F. Kamta*

## Abstract

Since its discovery in 2004, the graphene global market had a huge/considerable growth. Such growth can be explained by the use of graphene in specific or targeted applications where it has a huge and clear advantage. Although graphene is growing and has many possible applications, its market fraction is insignificant compared to the carbon global market. This is simply explained because the industry still has challenges related to quality, costs, reproducibility and safety. In this chapter, we propose a new look on the mechanical exfoliation. Basically, based on the difference in binding energy between graphite, graphene and a substrate we can exfoliate. The binding energy is the energy between materials at equilibrium. When 3 materials A-B-C are interacting, if the binding energy between A-B is superior to B-C, then by moving A in the opposite direction, B will follow. Based on that, we calculated the interaction potential between graphite, graphene and a substrate using the standard Lennard-Jones potential. Conventional substrates like silicon and silicon dioxide cannot exfoliate while gold, silver and copper can at 3.2 to 3.3 Å. This difference may be because of their higher atomic density and modest lattice parameter compared to others substrates used in this study.

**Keywords:** graphite, graphene, binding energy, cohesive energy, exfoliation

## 1. Introduction

Graphene, the world's thinnest material, was first introduced theoretically in 1947 by PR Wallace [1] to understand the properties of graphite. From this theoretical study, many scientific tried to isolate a single layer of graphite without success. It always resulted in few layers of graphene [2]. Graphene (single layer of graphite) becomes a reality in 2004 [3] when Andre Geim and Konstantin Novoselov isolated for the first time graphene using an adhesive scotch tape to exfoliate graphene and test its properties. The outstanding properties of graphene attracted a lot of interest in the scientific community and in the industry, opening a large range of opportunities and applications in electronics, medicine, etc. From 2004, the goal became then to find a reliable method to produce graphene layers. Today, we have many synthetics methods [4] that can be reliable in many aspects such as: Chemical Vapor Deposition (CVD), Liquid Phase Exfoliation (LPE) and reduction of graphene oxide. These methods allowed the transition from laboratory to industrial production.

In 2014, the global graphene market was projected to reach 190 million by 2022 [5], but recent studies by Report Link [6] shows that the graphene market will grow from 600 million USD (2020) to 1479 million in 2025. This huge grow in such a small time is really high, but compared to the carbon global market [7] (272 billion in 2020), it is almost insignificant. This is mostly because graphene is in its incubation phase and there is not yet a precise control on synthesis method. What could then be responsible for such growth? According to Li Lin and co-workers [8], the use of graphene in specific or targeted applications where it has a huge and clear advantage over other materials like in composites and electronics is the reason of graphene fast growth. More than that, the notable success of graphene in applications where chemical processes are required is because of its irregularities or defects [9]. This is to say that even though perfect flat graphene is hard to get, the small defective graphene plays a huge role in its growth.

Graphene has a large range of applications, but because of challenges related to quality, cost, reproducibility, processability and safety, we cannot really exploit it. So how, how to overcome these limits? Should we focus on enhancing the existing methods? Should we find new targeted applications? Should we focus more on standardization? Etc. The graphene industry still has a lot of work to accomplish, and anyone can work on something particular.

The aim of this project is to propose a new model of graphene exfoliation from Highly Oriented Pyrolytic Graphite (HOPG). Here, we gave a new look at the mechanical exfoliation by Andre Geim and Konstantin Novoselov [3]. Basically, using a scotch tape, they applied a force on top of graphite surface, graphite (many layers) sticks to the scotch, and by peeling it off using another scotch, they were able to get graphene. Why they did not get graphene the first time they put the scotch on graphite? Simply because the applied force was too superior to the interlayer force (cohesive or cleavage energy of graphite) that hold graphite layers together. Considering $F$ as the interlayer force of graphite, if the applied force $Fa$ was superior to $F$ and inferior to $2F$, then they would have got a single layer, but it is really difficult to achieve such small forces. Just like intermolecular forces like VDW forces are holding graphite layers together, putting a substrate to a certain distance from the graphite surface will create such forces. Using a substrate on top of graphite, our first goal is to show that we can exfoliate a single layer by controlling the resulting force (binding energy) between $F$ and $2F$. Thanks to the standard Lennard-Jones (LJ) potential we will calculate the binding energies of graphite surface/substrate and compare it to the interlayer binding energy of graphite (cleavage energy) to show that we can exfoliate. This work is divided into 3 parts. The first part is a review of exfoliation principles. The second part discusses our model and the calculations equations and finally the third part is about results and discussions.

## 2. Review on exfoliation principles and calculus

### 2.1 Exfoliation principles

Like gravitational forces acting on earth, there are intermolecular forces acting between all atoms and molecules. These forces have been variously known as dispersion forces, London forces, induced-dipoles forces, etc. but they are all referred as Van der Walls forces (VDW). As we know, these forces are present between graphite layers. In fact, the carbon layers are dipole-induced dipole that created forces between layers. These VDW forces created between two layers form a potential energy which is known as the bonding energy of graphite (binding energy). In order to exfoliate a graphene layer then, the cohesive energy has to be overcome. Exfoliation of graphite is a top-down method which consists in finding ways to surpass these interlayers forces, to breakdown or overcome that cohesive energy, hence the name of cleavage energy. We distinguish two types of exfoliation techniques for graphene: mechanical and chemical exfoliation.

In mechanical exfoliation a longitudinal or transversal stress is generated on the surface of the layered material by a scotch, AFM tip or a substrate [10]. The goal here is to overcome the energy between the layers. The difficulty in mechanical exfoliation is that the energy between layers is too small and by consequent easy to overcome resulting in many layers which is not graphene. Mechanical exfoliation is a simple and easy method for small lab experiments and cannot be used for industrial purpose. It is irreproducible, it has no control in layers, defects and size. Our goal is to change that, to make mechanical exfoliation reproducible and predictable by using intermolecular forces to exfoliate (**Figure 1**).

On the other hand, chemical exfoliation mostly known as liquid phase exfoliation (LPE) is almost the same as mechanical exfoliation. In LPE, transversal and longitudinal stress are provided by sonication, high shear mixing or micro fluidization [12] and it is happening in a liquid. LPE is a cheap, easy and scalable method to produce nano-flakes of graphite, but it has its challenges. Depending on the type of solvent used, we can face problems like aggregation, pollution, conservation and washing problems to some degrees. It also has a low yield (**Figure 2**).

## 2.2 Cohesive energy and implications

The first challenge for exfoliation technique is the requirement of an energy close to the cohesive energy of graphite, and we need to know its value first. In 1956, L. Girifalco and Lad [14], calculated the binding energy (cohesive energy) of graphite using the summation lattice with the LJ potential. They found the binding energy was approximately $-0.33\,J/m^2$ ($-53.96\,meV/atom$). They confirmed this result by comparing it to values got from Heat wetting experiment. Since the expansion of

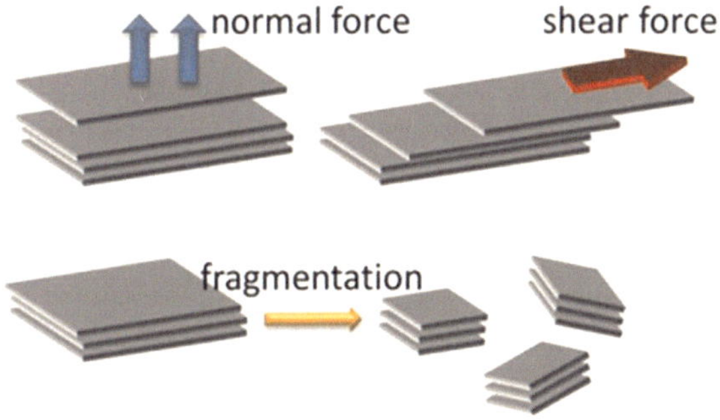


**Figure 1.**
*Mechanical exfoliation principle. (reprinted from reference Yi and Shen [11] with permission of CCC Inc.).*

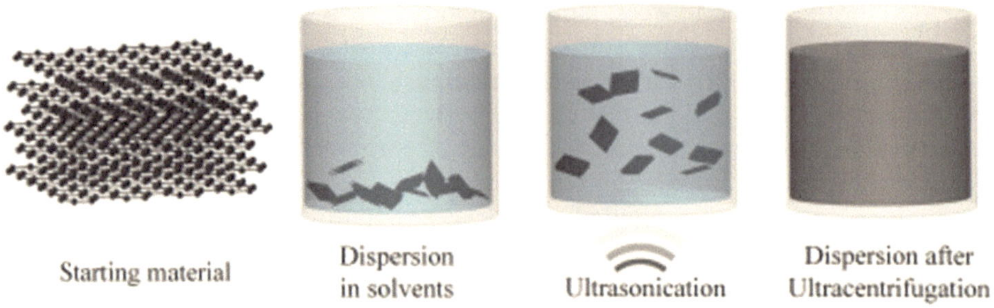


**Figure 2.**
*LPE using sonication (reprinted from reference [13]).*

graphene, the cohesive energy of graphite has attracted a lot of attention to understand better its properties for better exfoliation experiments design.

Xiaobin and colleagues calculated the interlayer potential by combining the Mobius inversion method with ab initio calculation [15]. They found the binding energy to be between 55 and 60 $meV/atom$. According to them, the binding/cohesive energy is between 50 and 60 $meV/atom$. They also found the interlayer space to be 3.1 Å. In the same way, using Quantum mechanics orbital occupancy approach and second order perturbation theory, Y.J Dappe and colleagues obtain a binding energy of 60–72 $meV/atom$ and interlayer space distance of 3.1–3.2 Å [16]. Throughout our literature review, we can say that the binding energy of graphite $Eg$ lies between 50 and 72 $meV/atom$ and the interlayer distance is 3.34 Å as confirmed by XRD studies of graphite. So the goal of all exfoliation technique is to provide an energy $Ex$ (by many means) which is: $Eg < Ex < 2 * Eg$ to exfoliate a single graphene layer. When $Ex$ exceeds, we can have bilayers and more.

## 2.3 Exfoliation methods

Using this knowledge about the cohesive energy, Zhang et al. [17] while working on transport measurements of microscopic graphite devices, fabricated their samples using a micromechanical cleavage process. With a silicon cantilever with a normal force between 10 – 2000nN (**Figure 3a**), they could exfoliate few layers of graphite. These small VDW forces between graphene layers (2 eV/nm$^2$) are not easy to achieve for graphene exfoliation but is required. For instance, Hernandez et al. [19] while working on LPE, found that by using a solvent with similar surface tension like graphene, the exfoliation was easier and single layers were obtained.

The requirement of such small forces made researchers to look for more sensible methods. For example, Seunguen Lee and Wei [18] simulate the exfoliation of graphene from HOPG on a SiO2 substrate and called it friction-induced exfoliation (**Figure 3b**). In that process, they used the lubrication properties of HOPG, the corrugation of HOPG combined with the roughness of the substrate. Another similar article is the exfoliation of graphene using the adhesive energy of few layers of graphene on a silicon substrate, while applying an upper velocity force on top of graphite to exfoliate the top layer (**Figure 3c**) [20]. In that study, based on the number of layer and the velocity, they could exfoliate the top layer.

In this paper, we present a new way of graphene exfoliation based on binding and interaction energy between graphene, graphite and a substrate.

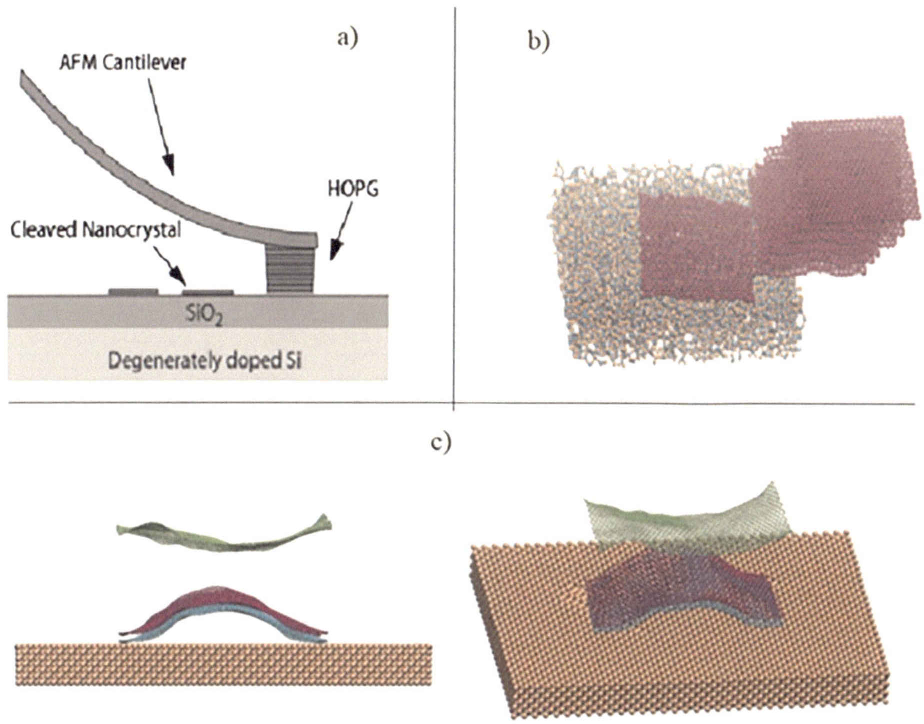


**Figure 3.**
*a. Cleavage process using the adhesion energy between graphite and substrate (reprinted from [17] with permission of AIP publishing) b. Friction induced exfoliation (reprinted from [18] with permission of AIP publishing) c. Top layer exfoliation based on velocity and adhesion.*

## 3. Proposed exfoliation and principle

### 3.1 Exfoliation model

When two materials (atoms, material made of atoms) are brought together, we can distinguish two types of energy: adhesive and interactive energy. The adhesive energy describes the forces when two materials are in contact (when they adhere) while the interaction energy describes the forces exerted by each material on the other. The interactive energy can be described by the Lennard-Jones (LJ) potential in **Figure 4**. When we look at the interactive energy curve, there is a minimum which is known as the binding energy $\varepsilon$. It is obtained at equilibrium. The broader and deeper it is, the stronger is the bond between the materials (mostly for non-covalent bonds). $\sigma$ is the distance at which the potential is 0. It can be taken as the repulsive core.

The working principle of our exfoliation model is shown in the next figure and can be viewed here (**Video 1**: https://drive.google.com/file/d/1eE48rBa0de_vzHd_zfxdQN940se1UqBj/view?usp=drivesdk). It is based on the interaction energy between three materials: graphite (HOPG), graphene (the bottom layer of HOPG) and a substrate. As we can see in the **Figure 5**, we consider HOPG as graphite and graphene. So our goal is to study the interaction between these materials.

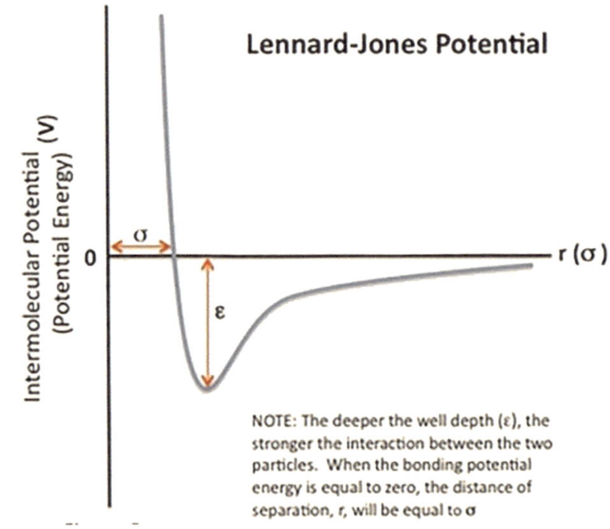


**Figure 4.**
*Lennard Jones potential curve.*

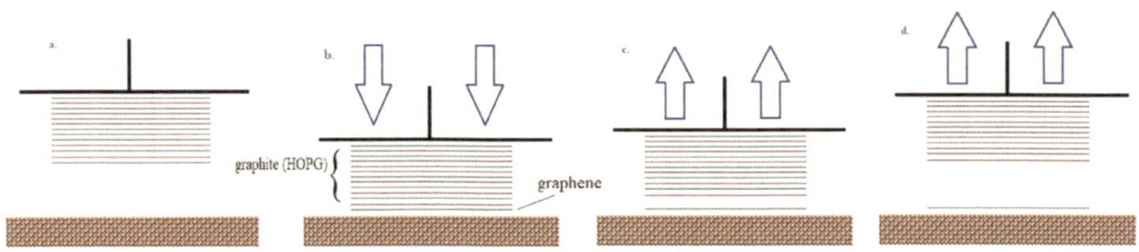


**Figure 5.**
*Exfoliation of graphene based on interaction energy.*

Here is a small description of the working principle.

a. The HOPG graphite and the substrate are separated considerably. The substrate is stable while the HOPG can move up and down.

b. The HOPG is pushed down to the substrate at a certain distance (the distance at which we have the binding energy between graphene and the substrate)

c. We pull up the HOPG and because the binding energy between the bottom layer graphene and the substrate is superior to the binding energy of graphite, the bottom layer will stay and adhere on the substrate.

d. Finally, the system is back to the normal and the HOPG is reduced by one layer and we can pass another substrate.

We simulated first the interaction energy (binding energy) between graphene and graphite. Then we determine the interaction energy between graphene and a couple of materials to see what materials can exfoliate and finally a study of the results to see how we can make the process more effective.

## 3.2 Calculation of the interaction energy using the LJ potential

To calculate the interaction energy between materials in this project, we used the standard LJ 6–12 potential expressed as:

$$V(r) = 4\varepsilon\left[\left(\frac{\sigma}{r}\right)^{12} - \left(\frac{\sigma}{r}\right)^{6}\right] \tag{1}$$

where $V(r)$ is the interaction energy between two atoms. The first term of the potential $\left(\frac{\sigma}{r}\right)^{12}$ is the repulsive term, and the second term $\left(\frac{\sigma}{r}\right)^{6}$ is the attractive term. $\varepsilon$ *and* $\sigma$ are the minimum value of the potential, and the distance at which the potential is zero respectively. $r$ is the distance between the atoms interacting. We opted for this potential because it is easy to manipulate, good for non-covalent bonds like VDW forces and more used for qualitative purpose.

### *3.2.1 Potential in terms of material*

The potential energy is the sum of a long range attractive contribution and short range repulsion as we saw in the potential curve. One thing about this LJ potential is that the bond strength increase with the atomic number/atomic mass (molecular weight). This is because atoms with large atomic numbers have high number of electrons that will then increase the VDW forces to get stronger. Second, the shape of the molecules also plays a role in the VDW bond strength. The shape will dictate the spread of the electrons, whether they are widely distributed or centered. For example, molecules of the same molecular weight have greater bond strength if they have larger aspect ratio shapes.

These understanding helps us in the designing or the choice of the materials that we are going to use to exfoliate graphene from graphite. From the previous paragraphs, we asked the following questions:

- What would be the material composition (atomic composition)?
- What could be the shape, size and aspect ratio?
- What should be the perfect material density to exfoliate just one graphene layers without causing defects or anything else?
- Is it possible to manufacture the material size that we want? Atomic thickness for example!

### *3.2.2 Parameters $\varepsilon$, $\sigma$*

The LJ potential is a pairwise potential. Atoms parameters of $\varepsilon$ *and* $\sigma$ are mostly defined by experiments. The parameters values $\varepsilon$,$\sigma$ for different atoms combinations were obtained through Rappe et al. [21], and the references [22, 23] by using the formulas:

$$\sigma_{A-B} = \frac{\sigma_{A-A} + \sigma_{B-B}}{2} \tag{2}$$

$$\varepsilon_{A-B} = \sqrt{\varepsilon_{A-A}\varepsilon_{B-B}} \tag{3}$$

| Pair_style | $\sigma(A)$ | $\varepsilon$(meV) |
|---|---|---|
| C – C | 3.431 | 4.55 |
| Si – Si | 3.826 | 17.4 |
| Ag – Ag | 2.648 | 350.6 |
| Au – Au | 2.646 | 412.9 |
| Cu – Cu | 2.338 | 409.32 |
| Si – C | 3.629 | 8.91 |
| O – C | 3.275 | 3.44 |
| Cu – C | 2.8845 | 43.156 |
| Au – C | 3.0385 | 43.3439 |
| Ag – C | 3.0395 | 39.9403 |

**Table 1.**
*Lennard Jones parameters of atoms and pair of atoms.*

**Table 1** shows the parameters values calculated for atoms combination we will use for simulations.

### *3.2.3 Computation process (pairwise summation)*

The LJ potential $V(r)$ as described in Eq. (1) is the interaction between two atoms a and b. When we talk about the interaction energy of graphene with a substrate, we are referring to more than two atoms (100, 1000 atoms for example). In general, to determine the interaction energy between two bodies G (graphene or graphite) and S (substrate), we do the summation of the LJ potential $V(r)$ over all the possible combinations of atoms in the bodies. For instance, the interaction energy between G and S is given by:

$$U(a) = \sum_{i}\sum_{j>i} V(r_{ij}) \tag{4}$$

where $r_{ij}$ is the separation distance between atoms i and j in the bodies G and S respectively. To compute the previous formula Eq. (4), we use the same method described by Giralfco and co-workers in their paper [14]:

- We calculate first the interaction between a carbon atom in graphene and all the atoms in the substrate. Since there is two type of carbon atoms in graphene, we have the following expressions

$$U_{1j}(a) = \sum_{j} V(r_{1j}) \tag{5}$$

$$U_{2j}(a) = \sum_{j} V(r_{2j}) \tag{6}$$

where $a$ is the distance between the graphene surface and the substrate surface as we can see in **Figure 6**; $r_{1j}$ and $r_{2j}$ are the distance between a carbon atom 1 (or 2) and the j atom in the substrate.

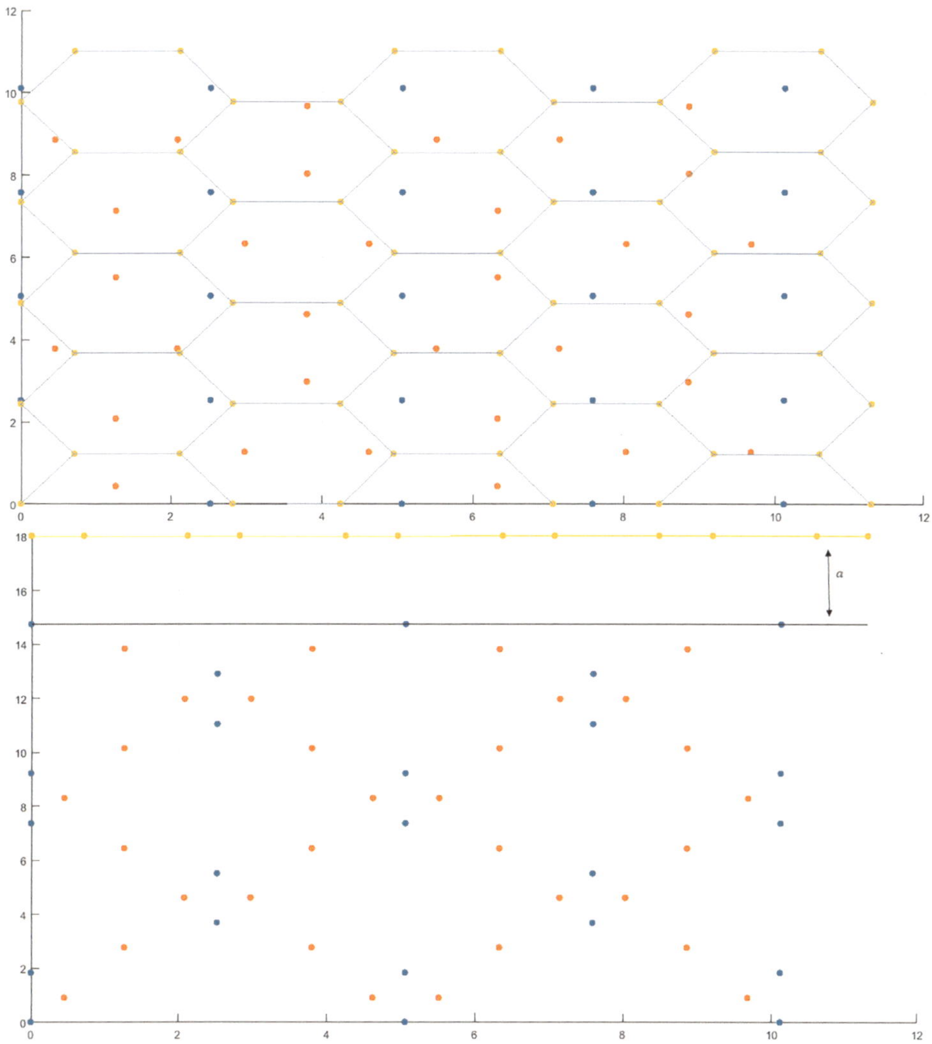


**Figure 6.**
*Left – xy view of graphene/silicon dioxide interaction; right – xz view of graphene/silicon dioxide separated. The yellow, blue and red points are carbon, silicon and oxygen atoms respectively.*

- After, to determine the interaction energy between a graphene sheet and the substrate, we integrate the previous terms Eqs. (5) and (6) over all the graphene sheet. Since here we are interested in general results, we are dividing the previous terms by the surface of the atoms ($\delta$: surface of one carbon atom).

$$U_{GS}(a) = \frac{1}{2\delta}\left(U_{1j}(a) + U_{1j}(a)\right) \tag{7}$$

where $\delta$ is the area occupied by a carbon atom in a graphene sheet. $\frac{1}{2\delta}$ is the number of atom of each type per square unit area in the monolayer.

### 3.2.4 Simulation set-up

For the simulation, we are using 2–3 supercells depending on the length of the lattice of the substrate. For example, silicon has a lattice parameter of $5.43^{\circ}A$. So, we are using two supercells for a total dimensions of 10.86*10.86*10.86. Based on those dimensions, we define the dimensions of graphite $11*11^{\circ}A^2$ for this case. We choose such a small set up to gain in simulation time and performance of the algorithm.

There are 5 simulations. Cupper, silicon, gold and silver have a face-centered cubic (fcc) lattice structure. So we will not show all the figures. By showing only one system we can guess how the other ones will look like since they have the same lattice structure with different lattice parameter. For instance, we have: Silicon ($5.43^{\circ}A$), Cupper ($3.625^{\circ}A$), Silver ($4.09^{\circ}A$) and Gold ($4.08^{\circ}A$).

The other simulation is with silicon dioxide (**Figure 6**). SiO2 is a complex structure that can be found in many structures. In this project, we consider silicon dioxide as a crystal in a simple tetragonal structure. Here the dimensions of the system are $10.5*10.5*14^{\circ}A^3$.

There are five simulations. The principle of our algorithm is to calculate the interaction energy as a function of the separation distance *a* between the substrate and the graphene/graphite.

## 4. Results and discussions

Most of the researches that has been done between graphene and a substrate is more concerned about the adhesive energy. Few papers are actually about the interaction energy or binding energy. For this reason, we first verified the acceptance of our algorithm by calculating the cohesive energy of graphite (cleavage energy) and compare the result to see if it is similar to others in the literature. From the computed interaction potential between graphite and graphene, we found the cohesive energy of graphite or the binding energy between graphite and graphene to be $-0.3060\ J/m^2$ at $3.6^{\circ}A$ (**Figure 7**). This result is closed to those got by Girifalco [14] and some practical experiments [24]. Now that we know our algorithm is acceptable, our goal is to compare the interaction between graphite – graphene and graphene – substrate. In the next figures, the blue curve represents the energy potential of graphene/graphite and the orange curve, the potential of graphene/substrate.

Using the same algorithm, we computed the interaction potential between silicon and graphene. We found the binding energy to be $-0.1014\ \mathrm{J/m^2}$. This result agrees with the one obtained by Norio Inui and Sho Iwasaki [22]. As we can see from the **Figure 8**, the cohesive energy of graphite is much more superior to the binding energy between graphene and silicon. From this energy difference, it is clear that silicon cannot exfoliate graphene from graphite.

For silicon dioxide, we found the binding energy with graphene to be $-0.090\ J/m^2$at $3.45^{\circ}A$ (**Figure 9**) which is the value got by Ishigma et al. [25] in their work. By comparing it to the cohesive energy of graphite, silicon dioxide cannot exfoliate graphene using this principle. However, a study by Wei Gao and coworkers shows that the interaction strength is strongly influenced by changes in the silicon dioxide surface structure due to surface reaction with water [26]. By reconstructing a $SiO_2$ surface, they could increase the interaction energy. In this project, we used silicon dioxide simple

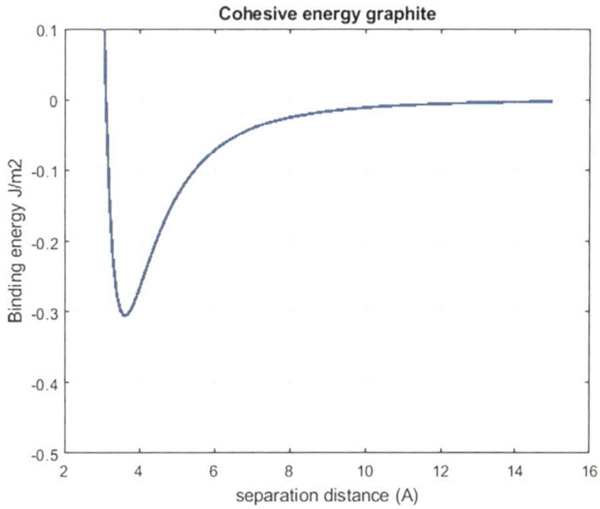


**Figure 7.**
*Graphite graphene interaction energy.*

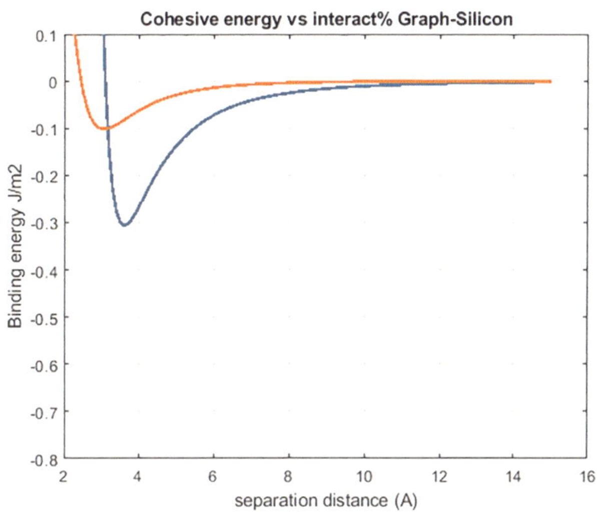


**Figure 8.**
*Graphene silicon interaction (orange), graphene/graphite interaction (blue).*

tetragonal structure without surface modification. This is to say that we can increase the binding energy by surface modification.

In the case of metals substrates, we find the binding energy between copper and graphene to be $-0.3792\, J/m^2$ at $1.2\mathring{A}$ (**Figure 10**). Here, we can notice that the equilibrium distance at which we obtain the binding energy is small. At first, it is tempting to say that copper cannot exfoliate graphene since at $1.2\mathring{A}$, the cohesive energy of graphite is infinity. But in practice, the interlayer distance between graphene layer is fixed, so by approaching graphite at $1.2\mathring{A}$ from the copper surface, we can see a weak exfoliation since they almost have the same energy. However, the distance is small.

The obtained binding energy between gold/graphene and silver/graphite is $-0.5924\, J/m^2$and $-0.5426\, J/m^2$ both at $3.2\mathring{A}$ respectively (**Figures 11** and **12**). They

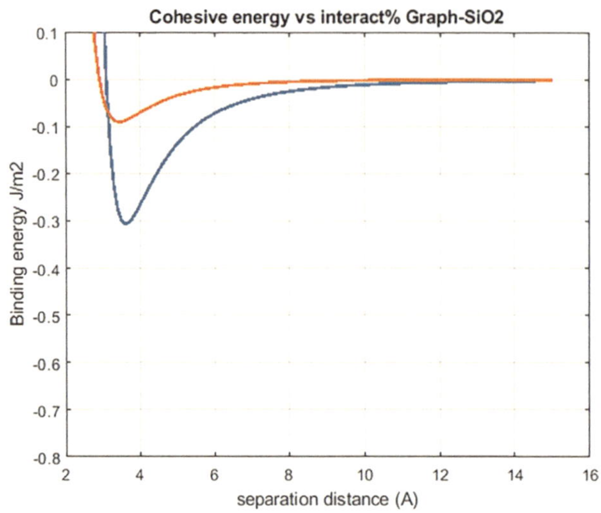


**Figure 9.**
*Graphene/silicon dioxide interaction potential (orange curve), graphite/graphene interaction (blue curve).*

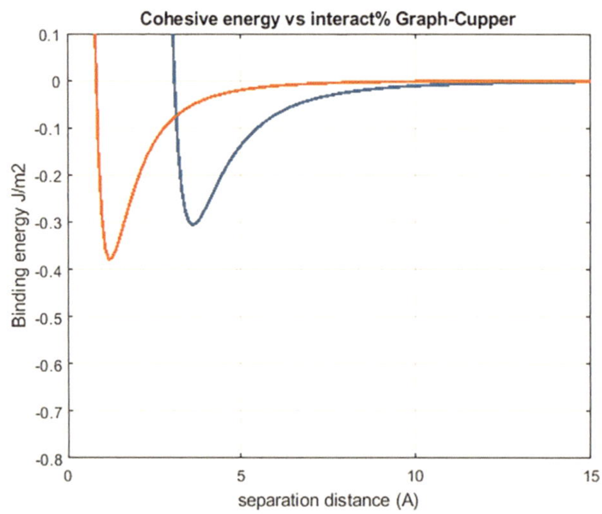


**Figure 10.**
*Graphene/cupper interaction potential (orange curve), graphite/graphene interaction (blue curve).*

both can clearly exfoliate graphene based on the principle described in **Figure 5**. These values of binding energy largely differ from copper and this may be because of the large atomic number and higher lattice parameter. From this result, we can agree that the atomic density has more influence on the binding energy.

## 5. Conclusion

The graphene industry still has a lot of rooms to improve in terms of standardization, conservation, synthesis, etc. The lack of a reliable and cost effective synthesis method avoids graphene to fulfill its potential. In this project, we proposed for the

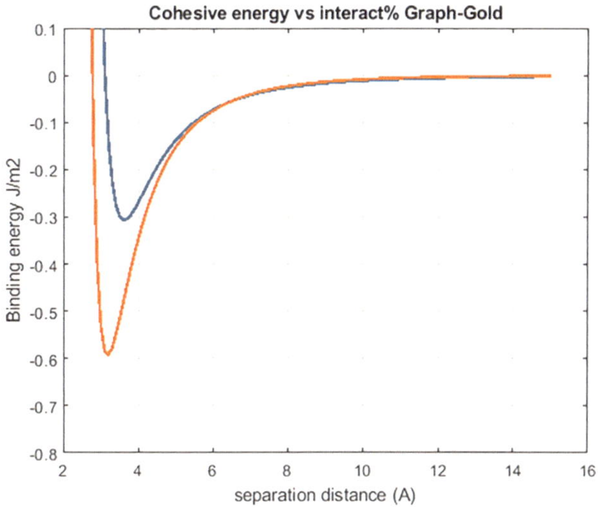


**Figure 11.**
*Graphene/gold interaction potential (orange curve), graphite/graphene interaction (blue curve).*

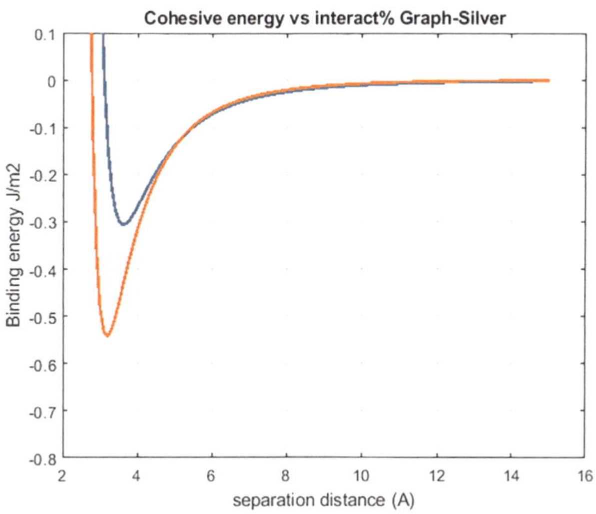


**Figure 12.**
*Graphene/silver interaction potential (orange curve), graphite/graphene interaction (blue curve).*

first time an exfoliation model of graphene from HOPG graphite based on the cohesive energy of graphite and the binding energy of the substrate with graphene. Calculations using the Lennard Jones potential and lattice summation showed metal substrate like gold, silver and copper can exfoliate graphene from graphite at 3.2 Å. Conventional substrate like silicon and silicon dioxide showed a binding energy inferior to the cohesive energy of graphite and cannot therefore exfoliate graphene. This result opens a new range of possibilities and opportunities for the graphene industry, notably in applications requiring first higher graphene quality. However further studies are still going on. First the distance at which the exfoliation can happen is small. 3.2 Å from the graphite or substrate surface is difficult to achieve. So the

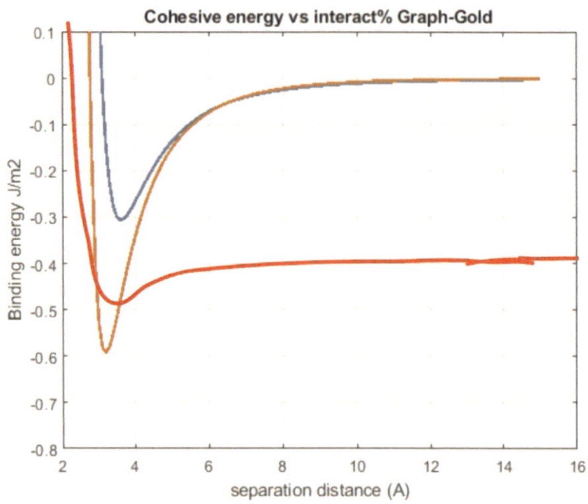


**Figure 13.**
*Perfect substrate behavior (red curve).*

challenge is to find or design a material that can exfoliate at an achievable distance which will make the process easier and effective. The perfect material curve looks like the one in **Figure 13**. where the red curve would be the behavior of the "perfect material". Second we need to find substrate materials that are polyvalent and can be used anywhere. Finally, we need to put more accuracy on our algorithm and do some practical tests. The truth is that this method is working, simply because by touching softly graphite or a pencil, there will be graphene layers in our finger and our goal was to understand it and automatized it. This model as shown in **Figure 5** can be easily scaled up to industrial manufacturing if we can achieve a good exfoliation distance. One thing is sure, a reliable, reproducible and effective synthesis technique could have a significant impact in the graphene industry.

## Acknowledgements

I whole-hearted express gratitude to my guide Dr. Ashwath Narayana for his guidance and teachings.

## Notes/thanks/other declarations

I want to thank God for blessing me with whatever I have.

Thanks to my parents who always support me.

A special thanks to my friends Draha Davy and Saadeu Jorick for their amazing friendship.

## Author details

Yannick A.F. Kamta
Rajiv Gandhi Institute of Technology (RGIT), Bangalore, India

*Address all correspondence to: kamtayannick@outlook.fr

## References


[1] Wallace PR. The band theory of graphite. Physical Review. 1947;**71**:622-634. DOI: 10.1103/PhysRev.71.622

[2] Boehm HP, Clauss A, Fischer G, Hofmann U. Surface properties of extremely thin graphite lamellae. In: Proceedings of the Fifth Conference on Carbon. Oxford, United Kingdom: Elsevier Inc; 1962. pp. 73-80. DOI: 10.1016/B978-0-08-009707-7.50013-3

[3] Novoselov KS, Geim AK, Morozov SV, Jiang D-e, Zhang Y, Dubonos SV, et al. Electric field effect in atomically thin carbon films. Science. 2004;**306**(5696):666-669. DOI: 10.1126/science.1102896

[4] Santhiran A, Iyngaran P, Abiman P, Kuganathan N. Graphene synthesis and its recent advances in applications—A review. C. 2021;**7**(4):76. DOI: 10.3390/c7040076

[5] Zion Market Research. Graphene Market (Mono-Layer & Bi-Layer Graphene, Few Layer Graphene, Graphene Oxide and Graphene Nano Platelets) for Composites, Energy Storage, Electronics and Others Applications: Global Industry Perspective, Comprehensive Analysis, and Forecast, 2016–2022. 2017. Available from: https://www.zionmarketresearch.com/report/graphene-market. [Accessed: February 17 2021]

[6] Report Linker. Graphene Market by Type, Application, End-use Industry and Region - Global Forecast to 2025. Available from: Graphene Market by Type, Application , End-use Industry and Region - Global Forecast to 2025 (reportlinker.com). [Accessed: March 28, 2021]

[7] Reuters. Global carbon markets value surged to record $277 billion last year – Refinitiv. 2021. Available from: Global carbon markets value surged to record $277 billion last year - Refinitiv | Reuters. [Accessed: February 18, 2021]

[8] Lin L, Peng H, Liu Z. Synthesis challenges for graphene industry. Nature Materials. 2019;**18**:520-524. DOI: 10.1038/s41563-019-0341-4

[9] Eftekhari A, Garcia H. The necessity of structural irregularities for the chemical applications of graphene. Materials Today Chemistry. 2017;**4**:1-16. DOI: 10.1016/j.mtchem.2017.02.003

[10] Das S, Choi W. Graphene: Synthesis and Applications. 1st ed. Boca Raton: CRC Press; 2011. DOI: 10.1201/b11259

[11] Yi M, Shen Z. A review on mechanical exfoliation for the scalable production of graphene. Journal of Materials Chemistry A. 2015;**3**(22): 11700-11715. DOI: 10.1039/C5TA00252D

[12] Xu Y, Cao H, Xue Y, Li B, Cai W. Liquid-phase exfoliation of graphene: An overview on exfoliation media, techniques, and challenges. Nanomaterials (Basel). 2018;**8**(11):942. DOI: 10.3390/nano8110942

[13] Hogan, Ben T., et al. 2D material liquid crystals for optoelectronics and photonics. Journal of Materials Chemistry C 543 (2017): 11185-11195. DOI: 10.1039/C7TC02549A.

[14] Girifalco LA, Lad RA. Energy of cohesion, compressibility, and the potential energy functions of the graphite system. The Journal of Chemical Physics. 1956;**25**:693-697. DOI: 10.1063/1.1743030

[15] Chen X, Tian F, Persson C, et al. Interlayer interactions in graphites. Scientific Reports. 2013;**3**:3046. DOI: 10.1038/srep03046

[16] Dappe YJ, Basanta MA, Flores F, Ortega J. Weak chemical interaction and van der Waals forces between graphene layers: A combined density functional and intermolecular perturbation theory approach. Physical Review B. 2006;**74**:205434. DOI: 10.1103/PhysRevB.74.205434

[17] Zhang Y, Small JP, Pontius WV, Kim P. Fabrication and electric-field-dependent transport measurements of mesoscopic graphite devices. Applied Physics Letters. 2005;**86**:073104. DOI: 10.1063/1.1862334

[18] Lee S, Wei L. Controlling the number of graphene sheets exfoliated from graphite by designed normal loading and frictional motion. Journal of Applied Physics. 2014;**116**:024313. DOI: 10.1063/1.4889924

[19] Hernandez Y, Nicolosi V, Lotya M, et al. High-yield production of graphene by liquid-phase exfoliation of graphite. Nature Nanotech. 2008;**3**:563-568. DOI: 10.1038/nnano.2008.215

[20] Zhang Q, Ma X, Zhao Y. Adhesion behavior between multilayer graphene and semiconductor substrates. Applied Sciences. 2018;**8**(11):2107. DOI: 10.3390/app8112107

[21] Rappe AK, Casewit CJ, Colwell KS, Goddard WA III, Skiff WM. UFF, a full periodic table force field for molecular mechanics and molecular dynamics simulations. Journal of the American Chemical Society. 1992;**114**(25): 10024-10035. DOI: 10.1021/ja00051a040

[22] Inui N, Iwasaki S. Interaction energy between graphene and a silicon substrate using pairwise summation of the Lennard-Jones potential. e-Journal of Surface Science and Nanotechnology. 2017;**15**:40-49. DOI: 10.1380/ejssnt.2017.40

[23] Ding S, Tian Y, Jiang Z, He X. Molecular dynamics simulation of joining process of Ag-Au nanowires and mechanical properties of the hybrid nanojoint. AIP Advances. 2015;**5**:057120. DOI: 10.1063/1.4921075

[24] Wang W, Dai S, Li X, et al. Measurement of the cleavage energy of graphite. Nature Communications. 2015; **6**:7853. DOI: 10.1038/ncomms8853

[25] Masa Ishigami JH, Chen WG, Cullen MS, Fuhrer, and E. D. Williams. Atomic structure of graphene on SiO2. Nano Letters. 2007;**7**(6):1643-1648. DOI: 10.1021/nl070613a

[26] Gao W, Xiao P, Henkelman G, Liechti KM, Huang R. Interfacial adhesion between graphene and silicon dioxide by density functional theory with van der Waals corrections. Journal of Physics D: Applied Physics; 2014;**47**: 255301. DOI: 10.1088/0022-3727/47/25/255301

Chapter 9

# Application of Electric Bias to Enhance the Sensitivity of Graphene-Based Surface Plasmon Resonance Sensors

*Ravi Paul Gollapalli, Tingyi Wei and Jeremy Reid*

## Abstract

Surface plasmon resonance sensors that incorporate graphene as one of the layers in the sensor structure have been proven to provide higher sensitivity in the detection of biomolecules, compared to sensors without graphene. Graphene an allotrope of carbon facilitates better adsorption to biomolecule samples due to the carbon-hydrocarbon affinity to biomolecules, thereby resulting in higher sensitive biosensors. Recently, a revolutionary method has been presented, at least in theory for now, that there is still a possibility to increase the sensitivity of the SPR sensors by the application of electric bias across the metal-graphene sensor system. A mathematical treatment to understand the physics of how the electrical bias contributes to an increase in sensitivity is presented in this chapter, using a sensor surface structure comprising of Au-MoS2-Gr. The results indicate that the application of electrical bias across the sensor surface consisting of Gr and other materials provides a method to increase the sensitivity of these biosensors. The scope and impact of this research can be felt in many industries that need sensors either in the food industry for food contamination check, harmful gas detection for environmental monitoring or safety measures, medical diagnostics etc.

**Keywords:** graphene, electric bias, enhanced sensitivity, surface plasmon resonance, biosensors

## 1. Introduction

Surface plasmon resonance (SPR) biosensors are based on the principle of surface plasmon generation, which is sensitive to the changes in the refractive index of the sensor surface structure along with the sensing medium. Surface plasmon (SP) is the collective oscillation of free electrons at the surface of a metal conduction band, which is excited by the electromagnetic field at the metal-dielectric interface. The SP exponentially decays into the surrounding media, and the decay depth is usually in the range of hundred nanometers, which is the key reason why the sensor surface of SPR-based sensors is in the nanometer range. Because the SP penetration depth is in the nanometer range makes them sensitive enough to detect changes in the refractive index in the milli to micro-range, making them the suitable and preferable sensors for detection of biomolecules in the micro-to-pico concentration levels. A highly sensitive biosensor is defined as one that can provide a measurable output signal for even the smallest changes in the refractive index of the sample under testing, and is often desirable.

### 1.1 Basic SPR sensor system

A basic (conventional) SPR sensor surface consists of a thin film, which interacts with the biomolecule material under testing also referred to as sensing medium. A change in the biomolecule concentration causes a local change in the refractive index near the metal surface of the SPR sensor, which results in a change in the propagation constant of the excitation optical wave. This change in the propagation constant can be optically measured [1]. Many different metals such as gold (Au), silver (Au), copper (Cu), aluminum (Al), etc. have been used to form the sensor surface and support the excitation of the surface plasmon usually dictates that the thickness of these metal films is in the 45–55 nm range. **Figure 1a** shows a typical SPR setup in a Kretschmann configuration. A beam of $p$-polarized laser light is incident on the metal-dielectric interface and the intensity of the reflected light is detected as a function of the angle of incidence. At a particular angle of incidence, called the surface plasmon resonance angle, $\theta_{spr}$ a resonance condition is satisfied and the surface plasmon is excited. The relation that governs the excitation of the surface plasmon is given as [2]:

$$\theta_{spr} = sin^{-1}\left(\frac{1}{n_p}\sqrt{\frac{n_m^2 n_d^2}{n_m^2 + n_d^2}}\right), \tag{1}$$

where $n_p$, $n_m$, and $n_d$ are the refractive indices of the prism, metal, and dielectric (sensing) medium, respectively. The excited surface plasmon is observed as a sharp drop in reflectance as shown in **Figure 1b**. As the refractive index of the sensing medium changes the resonance angle changes. SPR-based sensors essentially detect the shifts in the resonance angle for different materials under test. The "angular sensitivity" of an SPR setup is commonly defined as the ratio of the shift in the SPR angle, $\Delta\theta_{spr}$ to a given change in the refractive index, $\Delta n$ of the sensing medium and is given as [3]:

$$S_n = \frac{\Delta\theta_{spr}}{\Delta n}. \tag{2}$$

Along with the parameter, $S_n$, sensitivity, two other parameters figure of merit (FOM), and quality factor (QF) are used to evaluate the performance of an SPR system and are defined as [2]:

$$FOM = \frac{\Delta\theta_{spr}}{FWHM} \tag{3}$$

$$QF = \frac{S_n}{FWHM} \tag{4}$$

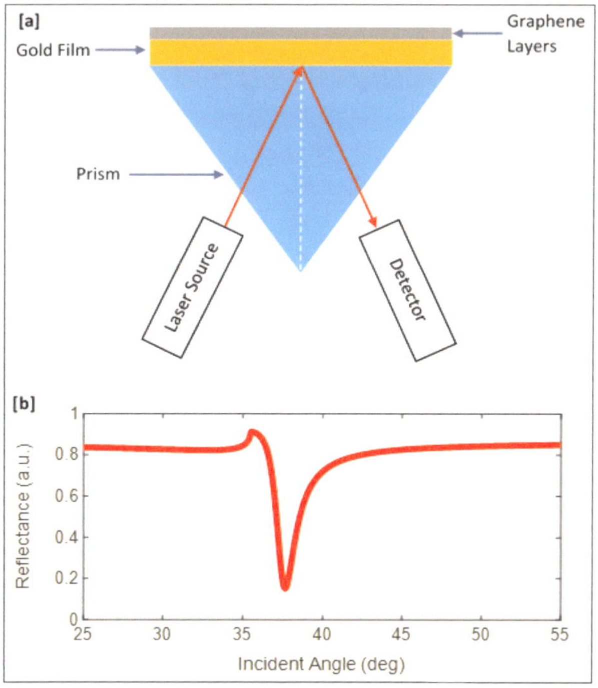


**Figure 1.**
*(a) A SPR system layout in Kretschmann configuration showing Au-Gr layers as the sensor surface. (b) A typical normalized reflectance curve showing the surface plasmon excitation as a drop in reflected intensity at 37.59 deg incident angle.*

## 1.2 SPR angle dependence on the refractive indices $n_p$, $n_m$, and $n_d$

$$\theta_{spr} = sin^{-1}\left(\frac{1}{n_p}\sqrt{\frac{n_d^2 n_m^2}{n_d^2 + n_m^2}}\right) \tag{5}$$

where, $n_p$, $n_d$, and $n_m$ are the refractive indices of the prism, dielectric (analyte), and metal. Given this relation, we can look at the variation of the resonance angle, with respect to the changes in the refractive index of the metal layer.

$$\frac{d\theta_{spr}}{dn_m} = \frac{n_d^3}{n_d^2 + n_m^2} \times \frac{1}{\sqrt{n_p^2\left(n_d^2 + n_m^2\right) - n_d^2 n_m^2}} \tag{6}$$

Given the above relation, we can see that the change in the resonance angle is a function of the refractive indices of the dielectric (sample under test), prism material, and also the metal. Now, we can also find the values of $n_p$, $n_d$, and $n_m$ for which the maximum resonance angle occurs when $d\theta_{SPR}/dn_m = 0$. This gives us two possible scenarios when either of the two terms in the equation can be zero.

### 1.2.1 Case #1

Consider the first term of Eq. (2),

$$\frac{n_d^3}{n_d^2 + n_m^2} = 0, \tag{7}$$

this can happen when

$$n_d = 0; n_d^2 + n_m^2 \to \infty, \tag{8}$$

the condition $n_d = 0$ dictates that the dielectric material (sample under testing) needs to have its refractive index to be zero, which is not a practical case. The condition, $n_d^2 + n_m^2 \to \infty$, combined with the earlier condition $n_d = 0$, dictates that:

$$n_d \to 0; n_m > > n_d; n_m \to \infty. \tag{9}$$

### 1.2.2 Case #2

Now, consider the second term of Eq. (2),

$$\frac{1}{\sqrt{n_p^2\left(n_d^2 + n_m^2\right) - n_d^2 n_m^2}} = 0, \tag{10}$$

this term shows that the maximum resonance angle occurs when, $\sqrt{n_p^2\left(n_d^2 + n_m^2\right) - n_d^2 n_m^2} \to \infty$. This stipulates that to achieve maximum resonance angle, the product $n_p^2\left(n_d^2 + n_m^2\right) \to \infty$ while $n_d^2 n_m^2 \to 0$. These two conditions combined with the conditions in Eq. (5) indicate that the most favorable conditions to achieve the maximum plasmon resonance angle for better sensitivity are:

$$n_d \to 0; n_m \to \infty; n_p \to \infty. \tag{11}$$

Up till now, we have used $n_m$ to represent the refractive index of the metal layer used to excite the surface plasmon, however, there can be many possible combinations of multi-layers of different materials, and the $n_m$ would be the effective refractive index of all these combined multi-layers used to excite the surface plasmon. As has been the case in the past decades where many researchers have proposed varied types of sensor surfaces with combinations of Au, Ag, Al, Si, $SiO_2$, $MoS_2$, Black Phosphorus (BP), etc., with varying thicknesses optimized for high sensitivity.

## 1.3 Ultra-sensitive SPR sensor design

Many researchers have proposed and shown that other metals, or combinations thereof, can be used to improve the sensitivity of the SPR system. Many methods have been demonstrated to improve the SPR biosensor sensitivity, such as, utilizing metal nanoparticles and nanoholes [4], metallic nanoslits [5], and colloidal gold nanoparticles [6] in buffered solutions. In the conventional SPR system, gold is commonly preferred due to its resistance to oxidation and corrosion, however, biomolecules adsorb poorly on gold, causing less interaction with the metallic surface and this limits the sensitivity of the SPR biosensor. One of the attractive alternatives to improve sensitivity by increasing the adsorption of biomolecules on the gold surface is to use graphene (Gr). Among these materials, graphene (Gr) due to its carbon composition has proven to be a more preferable material due to its affinity and absorptive nature to hydrocarbons. This not only helped in the binding process of biomolecules under testing but also helped improve the sensitivity compared to the conventional system [7, 8]. Gold-graphene SPR biosensors have been shown to stably adsorb biomolecules with carbon-based ring structures, causing an increase in sensitivity to changes in refractive index of the biomolecules, leading to increased detection and identification [9, 10]. However, the precise control over the geometry and the optical properties of nanostructures is challenging. Many methods were demonstrated to improve the sensitivity, by either utilizing metal nanoparticles and nanoholes [4], metallic nanoslits [5], and colloidal gold nanoparticles [6] in buffered solutions. However, still improvements can be made using other techniques or methods. Recently, we proposed an idea to increase an SPR biosensor sensitivity by changing the refractive index of the graphene layer, which is a part of the sensor surface by the application of electrical bias voltage across metal-graphene system [11]. Our numerical investigation has shown that apart from providing an increase in sensitivity compared to other techniques and sensor surface structures, we also realized that this technique can also address another issue: control over the optical properties of the sensor surface [12].

### 1.4 Novelty and necessity of the present idea

In a typical SPR sensor setup, for a particular type of sensing medium, a particular sensor surface configuration is used. This means that for a different type of sensing medium and for an enhanced sensitivity, a different sensor surface may be required. This is combined with the fact that once a sensor surface configuration for use in the sensor is selected, it would not be possible to make changes to it, and the only option would be to switch sensor surface with another one. For example, if a gold-graphene sensor surface with certain thickness combination is chosen as sensor surface, but a particular sensing medium requires a different thickness combination that could result in better sensor sensitivity output, then the only option is to switch out the sensor surface. The reason for such a need to switch the sensor surface film is based on the fact that the effective refractive index of the sensor surface is constant for a particular wavelength of the plasmon excitation light source. We believe that if the sensor surface properties could be tuned dynamically, we could eliminate the need for a different sensor surface combination as would be required for different sensing media. As mentioned earlier, we have already presented our numerical results on the application of electrical bias across gold-graphene sensor surface structure in Ref [11], and gold-$SiO_2$-graphene in Ref [12], and here we extend our numerical investigation to the Au-MoS2-Gr sensor surface structure.

## 2. Electrical bias across the metal-Gr system

Graphene is a thin two-dimensional layer of graphite, an allotrope of carbon, where the carbon atoms are arranged in a honeycomb lattice structure. The valence and conduction bands in graphene touch at six points known as the Dirac points, where the chemical potential of graphene is located for undoped samples. The position of the chemical potential can easily be shifted above or below the Dirac point, thus allowing the carrier concentration in the material to be tuned by applying an electrical voltage [13]. In this Au-Gr system, the carrier concentration in graphene, $n_g$ due to an applied voltage, $V_g$ can be calculated using the relation [14]:

$$n_g = \frac{V_g \varepsilon_0 \varepsilon_r}{q d_{sub}} \tag{12}$$

where $\varepsilon_0 = 8.85 \times 10^{-12} F/m$ is the permittivity of vacuum, $\varepsilon_r$ is the relative permittivity of the substrate (50 nm thick gold film), $q$ is the electron charge, and $d_{sub}$ is the substrate thickness. Based on the carrier concentration of the system, the chemical potential $\mu_c$ can be calculated using [15]:

$$\mu_c = \hbar v_f \sqrt{\pi n_g} \tag{13}$$

where $\hbar$ is the reduced Planck's constant and $v_F \approx 10^6$ cm/s is the Fermi velocity. Graphene optical conductivity, $\sigma$ is related to the intra-band electron-photon scattering, $\sigma_{intra}$ and the inter-band electron transition conductivity, $\sigma_{inter}$ as function of radiation frequency, $\omega$ and given as:

$$\sigma(\omega) = \sigma_{intra}(\omega) + \sigma_{inter}(\omega) \tag{14}$$

which can be calculated using the Kubo formula [16]:

$$\sigma_{intra}(\omega) = i \frac{q^2}{\pi\hbar(\omega + i\tau^{-1})} \left[ \mu_c + 2K_B T \times ln \left\{ e^{\left(\frac{-\mu_c}{K_B T}\right)} + 1 \right\} \right] \tag{15}$$

$$\sigma_{inter}(\omega) = i \frac{q^2}{4\pi\hbar} ln \left[ \frac{2|\mu_c| - \hbar(\omega + i\tau^{-1})}{2|\mu_c| + \hbar(\omega + i\tau^{-1})} \right] \tag{16}$$

where $K_B$ is the Boltzmann's constant, $T$ is the temperature, $\tau = \mu_c m_u / q v_F^2$ is the momentum relaxation time, and $m_u = 10^4 cm^2/V - s$ is the impurity-limited direct current mobility.

The complex conductivity of graphene may be expressed as:

$$\sigma(\omega) = \sigma_R(\omega) + i\sigma_I(\omega) \tag{17}$$

where $\sigma_R(\omega)$ and $\sigma_I(\omega)$ are the real and imaginary parts of the graphene conductivity and can be derived from Eqs. (5)–(7) and given as:

$$\sigma_R(\omega) = \frac{\tau^{-1} q^2}{(\omega^2 + \tau^{-2})\pi\hbar^2} \times \left[ \mu_c + 2k_B T \times ln \left\{ e^{\left(\frac{-\mu_c}{k_B T}\right)} + 1 \right\} \right], \tag{18}$$

$$\sigma_I(\omega) = \frac{\omega q^2}{(\omega^2 + \tau^{-2})\pi\hbar^2} \times \left[ \mu_c + 2k_B T \times ln \left\{ e^{\left(\frac{-\mu_c}{k_B T}\right)} + 1 \right\} \right] + \frac{q^2}{4\pi\hbar} ln \left[ \frac{2|\mu_c| - \hbar(\omega + i\tau^{-1})}{2|\mu_c| + \hbar(\omega + i\tau^{-1})} \right]. \tag{19}$$

The thickness of a single graphene layer is 0.34 nm, and for a given number of graphene layers with thickness, $d_{Gr}$ the real $n_{Gr,R}$ and imaginary $n_{Gr,I}$ parts of the graphene refractive index can be expressed as:

$$n_{Gr,R} = \sqrt{\frac{\sqrt{(\sigma_I - \omega\varepsilon_0 d_{Gr})^2 + \sigma_R^2} - (\sigma_I - \omega\varepsilon_0 d_{Gr})}{2\omega\varepsilon_0 d_{Gr}}} \tag{20}$$

$$n_{Gr,I} = \sqrt{\frac{\sqrt{(\sigma_I - \omega\varepsilon_0 d_{Gr})^2 + \sigma_R^2} + (\sigma_I - \omega\varepsilon_0 d_{Gr})}{2\omega\varepsilon_0 d_{Gr}}} \tag{21}$$

From Eqs. (3)–(13), it is evident that the refractive index of the metal-graphene sensor surface system can be varied by the application of an electrical bias voltage. Using the N-layer model [1], the reflectance of the incident light in the Kretschmann configuration of SPR setup was calculated for this Au-Gr sensor surface system. **Figure 2** shows a top-view kind of perspective to how the SPR angle shifts at different levels of chemical potential. Using chemical potential as the basis to evaluate the shift in the SPR angle for a change in the sample refractive index, we defined a new measure of sensitivity $S_{\mu_c}$, which is the ratio of the change in the SPR angle due to a change in the chemical potential applied across the metal-graphene system [11],

$$S_{\mu_c} = \frac{\Delta\theta_{spr,\mu_c}}{\Delta\mu_c} \tag{22}$$

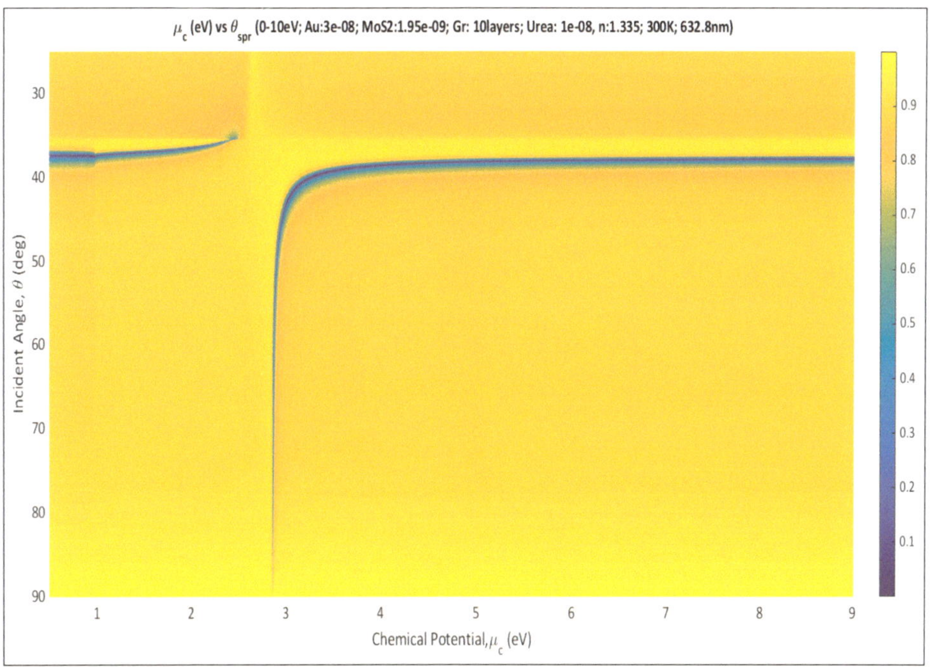


**Figure 2.**
*A top view kind of perspective showing how the SPR curves are changing wrt the chemical potential. With prism material as SF10 ($n_p$ = 1.723), the sensor surface configuration is: Au (30 nm thick), MoS2 (3 layers), Gr (10 layers) with $\mu_c$ at 0.0 eV, and the incident radiation wavelength $\lambda$ = 632.8 nm.*

along with the figure of merit wrt chemical potential as:

$$FOM_{\mu_c} = \frac{\Delta\theta_{spr,\mu_c}}{FWHM} \tag{23}$$

## 2.1 Modeling of the multi-material sensor surface structure

In this section let us look at how we can model the presence of more than one material such that it can be biased using an electrical voltage. To accomplish this, we can still consider the Au film as the base on which the graphene layers are deposited, as discussed in the Section 2, and the other material layers may be considered to be positioned either between the Au and Gr layers or exterior to the Au and Gr layers, and the following set of equations may be used for any novel sensor structures. **Figure 3** shows such a layout where MoS2 layer(s) are sandwiched between the Au and Gr layers. We will use this configuration to investigate the effect of electrical bias on the sensitivity of an SPR system.

MoS2 is a semiconductor material with an ultrathin direct bandgap and belongs to the transition metal dichalcogenide group, with characteristics similar to graphene. A strong coupling can be induced at the metal/MoS2 interface because of the effective charge transfer and large field enhancement resulting in improved SPR sensitivity [8, 17]. The nominal thickness of a single layer of MoS2 is 0.65 nm.

Consider the relation for capacitance of a capacitor:

$$C = \frac{\varepsilon_0 \varepsilon_r A}{d} = \frac{Q}{V}, \tag{24}$$

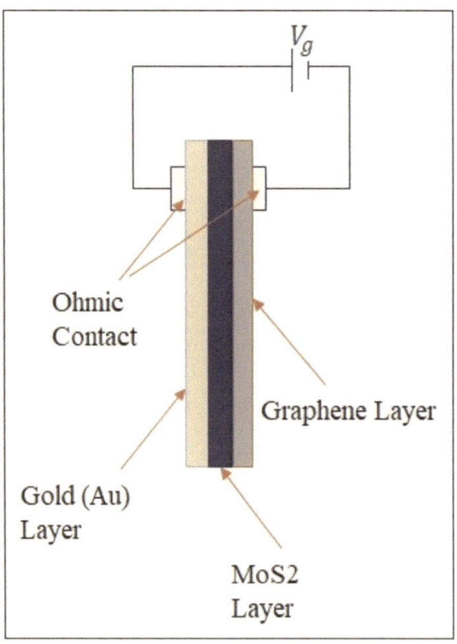


**Figure 3.**
*Diagram showing the three layers of the sensor surface and biased using ohmic contacts to a voltage source. Shown here are Au-MoS2-Gr layers forming, which can be modeled as capacitors in series. Note that the thicknesses of the layers shown are not to scale.*

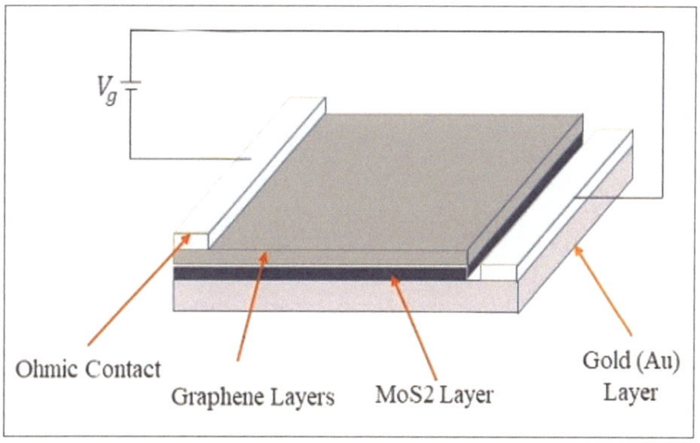


**Figure 4.**
*Diagram showing the three layers of the sensor surface and biased using ohmic contacts to a voltage source. Note that the thicknesses of the layers shown are not to scale.*

where $C$, $A$, and $d$ are the capacitance, surface area, and the separation distance between the plates of a capacitor, respectively. Modeling the Au-MoS2-Gr layers as two capacitors in series (**Figure 4**, where the equivalent capacitance of such system is:

$$\frac{1}{C_{eq}} = \frac{1}{C_1} + \frac{1}{C_2} \tag{25}$$

We can find the carrier concentration, $n_g$ in Gr on top of this Au-MoS2 layers, due to an applied voltage, $V_g$ given by:

$$n_g = \frac{V_g}{q} \frac{\varepsilon_0}{\left(\frac{d_{MoS2}}{\varepsilon_{r_{MoS2}}} + \frac{d_{Gr}}{\varepsilon_{r_{Gr}}}\right)} \tag{26}$$

where $\varepsilon_0 = 8.85 \times 10^{-12}$ F/m is the permittivity of vacuum, $q$ is the electron charge, $d_{MoS2}$, $d_{Gr}$, $\varepsilon_{r_{MoS2}}$, and $\varepsilon_{r_{Gr}}$ are the thickness and relative permittivity of MoS2 and Gr layers respectively.

Eq. (26) combined with Eqs. (13)–(21) can be used to calculate the refractive index of Gr as the electric bias $V_g$ is varied from 0 to 10 eV. The SPR resonant angle and reflection values can be found using relations given in Ref [1].

## 3. Results and discussion

**Figure 5** shows the SPR curves as the chemical potential is varied from 0 to 8 eV, for the case of sensor configuration with prism material, MgF2 ($n_p = 1.377$), followed by Au layer (30 nm thickness, $n_{Au} = 0.1726 + 3.4218$ i), MoS2 (1 layer, $n_{MoS2} = 4.8046 + 0.84395$ i, [18]), Gr (10 layers), and the sample considered is urea (10 nm thickness, $n_{Urea} = 1.335$), the wavelength used for these calculations is $\lambda = 632.8$ nm. To better understand the impact of the applied electrical bias across the Au-MoS2-Gr system and the shift in the SPR angle with respect to change in the refractive index of the urea sample, the SPR curves are referenced to the SPR curve when the urea sample has refractive index value n = 1.332. We will base our discussion, using $n_{urea} = 1.332$ as our reference, while all other values of the sensor surface material are the same. In the

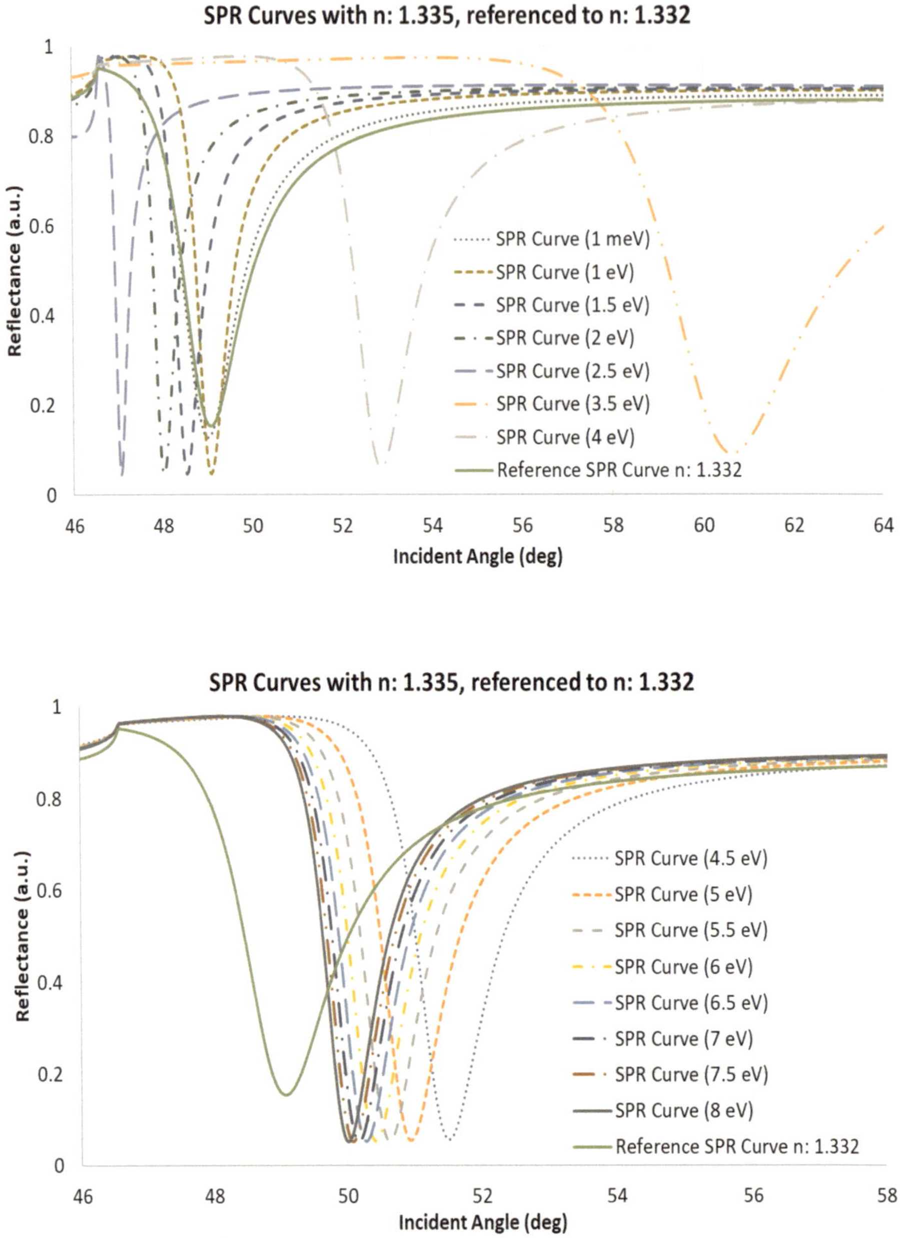


**Figure 5.**
*SPR Curves for the sensor configuration: MgF2 ($n_p = 1.377$), followed by Au layer (30 nm thickness, $n_{Au} = 0.1726 + 3.4218\,i$), MoS2 (1 layer, $n_{MoS2} = 4.8046 + 0.84395\,i$, [18]), Gr (10 layers), and the sample considered is urea (10 nm thickness, $n_{urea} = 1.335$).*

**Figure 5**, we can see that the value of the SPR dip, Rmin = 0.15 with SPR angle = 49.1 deg. for n = 1.332 and for n = 1.335, the Rmin = 0.13 with SPR angle = 48.9 deg., both at 1 meV applied chemical potential, which again will be our reference applied chemical potential value for the rest of our discussion of the results throughout this text. As seen in **Figure 2**, we also see in **Figure 5**, that the SPR curves move toward lower SPR angle from 49.1 deg. to 47.06 deg. at 2.5 eV; and then show up at the extreme right (90 deg) mark and move to the lower angles before settling down between 51.49 deg. at 4.5 eV to 50.1 deg. at 8 eV and it stays almost the same even as the chemical potential is increased further till 10 eV. We also see that the Rmin for all these $\mu_c$ values from 0.001 eV to 8 eV is below 0.05 (a.u.), which shows that there is a strong excitation of surface plasmon facilitated by the application of electrical bias. A shift of more than 1 deg. in the SPR angle is a very strong indication of the high sensitivity this technique provides as the sample changes its refractive index value from n = 1.332 to 1.335, which gives a sensitivity value of 333.33°/$RIU$. It should be noted that optimized graphene-based sensor surface structure for a particular bio-molecule can be further optimized for higher sensitivity by application electrical bias and without any changes to the sensor structure.

**Figure 6** shows the variation of the SPR angle as the chemical potential varied. It has already been shown that the number of graphene layers usually results in enhancement of the electrical, optical, chemical, and mechanical properties of device [19], but here we specifically want to highlight the impact of graphene layers on the SPR angle shift, which is what is done in this figure. This figure shows that when Gr layers = 12, at $\mu_c = 3$ eV, there is a very large shift of 28.91 deg. (from 49.46 at 1 meV to 20.55 deg. at 3 eV) whereas the SPR angle shift is 9.22 deg. and 7.48 deg. for Gr = 8 & 10 layer structure (refer to **Table 1**), which in their own right they are much larger than other optimized sensor structures reported for urea detection.

**Figure** 7 shows the variation of the Rmin and FWHM of the SPR curves with respect to the variation of chemical potential across the sensor surface. We see that the Rmin = 0.9 and FWHM = 16.05 deg. for Gr = 12 layers at $\mu_c = 3$ eV, which eliminates this as a not least favorable case for SPR measurement, and so we will have to choose the values obtainable at $\mu_c = 2.5$ eV, which provides Rmin = 0.04 and FWHM = 0.46 deg. However, if compared to the case of Gr = 8 and Gr = 10 layers we see that these two cases are far more favorable for SPR measurement because of their

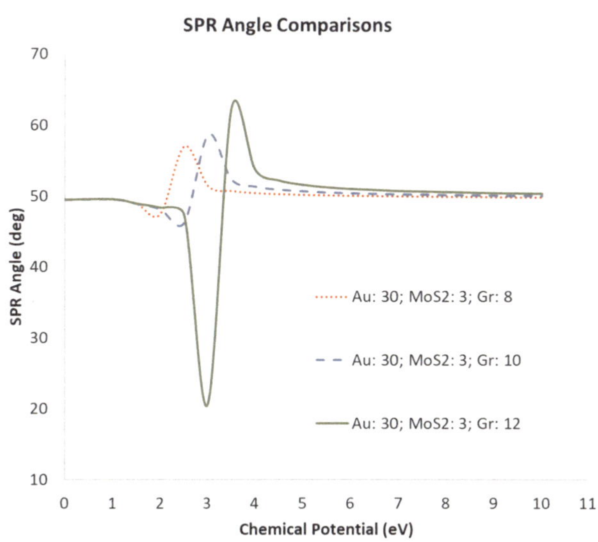


**Figure 6.**
*SPR Curves for the configuration: MgF2 ($n_p = 1.377$), followed by Au layer (30 nm thickness, $n_{Au} = 0.1726 + 3.4218\,i$), MoS2 (1 layer, $n_{MoS2} = 4.8046 + 0.84395\,i$, [18]), Gr (10 layers), and the sample considered is urea (10 nm thickness, $n_{urea} = 1.335$).*

| Gr layers → | 8 | 8 | 8 | 10 | 10 | 10 | 12 | 12 | 12 |
|---|---|---|---|---|---|---|---|---|---|
| $\mu_c$ (eV) ↓ | $\theta_{spr}$ | Rmin | FWHM | $\theta_{spr}$ | Rmin | FWHM | $\theta_{spr}$ | Rmin | FWHM |
| 0.001 | 49.50 | 0.070 | 1.69 | 49.46 | 0.095 | 1.76 | 49.46 | 0.112 | 1.87 |
| 1.000 | 49.50 | 0.048 | 1.04 | 49.53 | 0.048 | 1.08 | 49.53 | 0.047 | 1.04 |
| 1.500 | 48.88 | 0.047 | 0.90 | 48.92 | 0.047 | 0.90 | 48.92 | 0.048 | 0.90 |
| 2.000 | 47.44 | 0.400 | 0.54 | 48.13 | 0.044 | 0.72 | 48.34 | 0.046 | 0.75 |
| 2.500 | 56.98 | 0.070 | 3.06 | 46.26 | 0.728 | 4.17 | 47.26 | 0.043 | 0.46 |
| 3.000 | 51.37 | 0.055 | 1.47 | 58.68 | 0.080 | 3.63 | 20.55 | 0.908 | 16.05 |
| 3.500 | 50.68 | 0.053 | 1.29 | 52.34 | 0.058 | 1.65 | 62.42 | 0.100 | 5.25 |
| 4.000 | 50.40 | 0.052 | 1.22 | 51.33 | 0.055 | 1.44 | 53.67 | 0.063 | 1.98 |
| 4.500 | 50.25 | 0.052 | 1.18 | 50.90 | 0.053 | 1.33 | 52.16 | 0.058 | 1.62 |
| 5.000 | 50.14 | 0.051 | 1.15 | 50.68 | 0.053 | 1.26 | 51.55 | 0.056 | 1.47 |
| 5.500 | 50.07 | 0.051 | 1.11 | 50.50 | 0.053 | 1.22 | 51.19 | 0.054 | 1.36 |
| 6.000 | 50.00 | 0.051 | 1.11 | 50.40 | 0.052 | 1.22 | 50.97 | 0.054 | 1.33 |
| 6.500 | 49.96 | 0.050 | 1.11 | 50.32 | 0.051 | 1.18 | 50.83 | 0.054 | 1.26 |
| 7.000 | 49.93 | 0.050 | 1.08 | 50.25 | 0.051 | 1.15 | 50.68 | 0.053 | 1.26 |
| 7.500 | 49.89 | 0.050 | 1.08 | 50.22 | 0.052 | 1.15 | 50.61 | 0.053 | 1.26 |
| 8.000 | 49.86 | 0.051 | 1.04 | 50.14 | 0.052 | 1.15 | 50.54 | 0.053 | 1.18 |
| 8.500 | 49.86 | 0.051 | 1.08 | 50.11 | 0.051 | 1.11 | 50.47 | 0.052 | 1.18 |
| 9.000 | 49.82 | 0.050 | 1.04 | 50.07 | 0.051 | 1.11 | 50.40 | 0.052 | 1.15 |
| 9.500 | 49.78 | 0.051 | 1.04 | 50.04 | 0.051 | 1.08 | 50.36 | 0.052 | 1.18 |
| 10.00 | 49.78 | 0.050 | 1.00 | 50.04 | 0.051 | 1.08 | 50.32 | 0.052 | 1.15 |

*The maximum value of the SPR angle achievable, along with the corresponding Rmin and FWHM, are highlighted in red color. These maximum SPR angle values occur at 2.5, 3.0 and 3.5 eV for the case of 8, 10 and 12 Gr layers in the sensor system, which also indicates that maximum SPR angle occurs at higher applied electrical potential values as the Gr layers increase. The SPR angle, Rmin and FWHM values are highlighted in blue color as comparison with the values for the Au (50 nm) - Gr sensor system as described in [11].*

**Table 1.**
*Chemical potential (eV) versus SPR angle (deg), Rmin (a.u.), and the FWHM (deg) for different graphene layers for the sensor structure Au (30 nm) - MoS2 (3 layers) - Gr, with the urea as sample (thickness 10 nm, n = 1.335).*

low Rmin and FWHM when the maximum SPR angle shift occurs. So, one cannot only use the SPR angle shift as the sole parameter to understand the sensitivity of the SPR sensors, and so it is common to use FOM and QF to better evaluate the sensitivity of the SPR sensors.

**Figure 8** shows the variation of Sn and FOM for three different cases of Gr = 8, 10 &12 layers while Au-MoS2 thicknesses are 30 nm and 3 layers, respectively. The curves shown, here can be seen in **Table 2**. **Figure 9** shows the variation of $S_{n,\mu c}$ and $FOM_{\mu c}$ for Gr = 8, 10 &12 layers for the same Au-MoS2 thicknesses of 30 nm and 3 layers, respectively.

**Table 2** provides the values of Sn and FOM for varying chemical potential values using urea sample with n = 1.332 as reference and comparing the values to urea sample with n = 1.335. Also, note that the negative values of the Sn and FOM indicate the fact

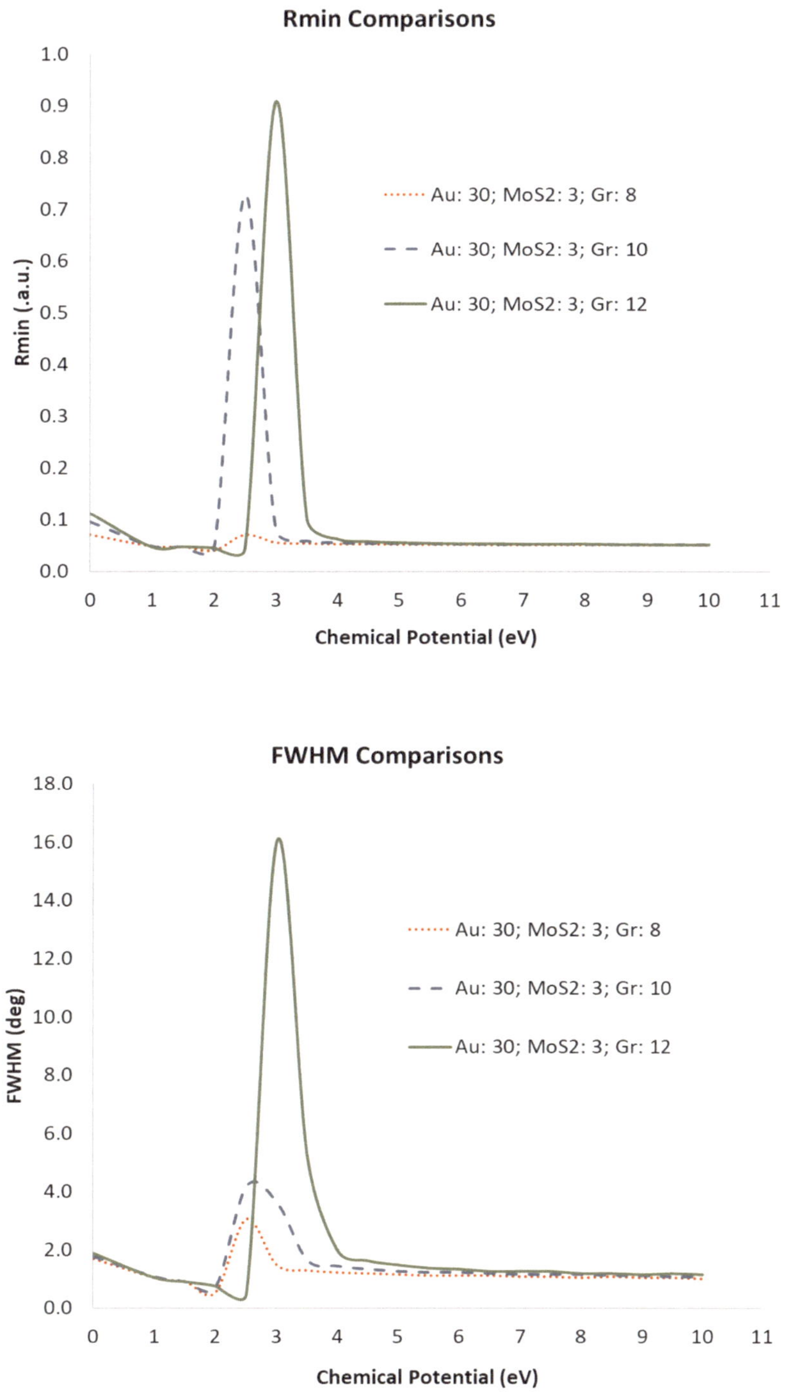


**Figure 7.**
*$R_{min}$ and FWHM curves for the sensor-sample: MgF2 ($n_p = 1.377$), followed by Au layer (30 nm thickness, $n_{Au} = 0.1726 + 3.4218\ i$), MoS2 (1 layer, $n_{MoS2} = 4.8046 + 0.84395\ i$, [18]), Gr (10 layers), and the sample considered is urea (10 nm thickness, $n_{urea} = 1.335$).*

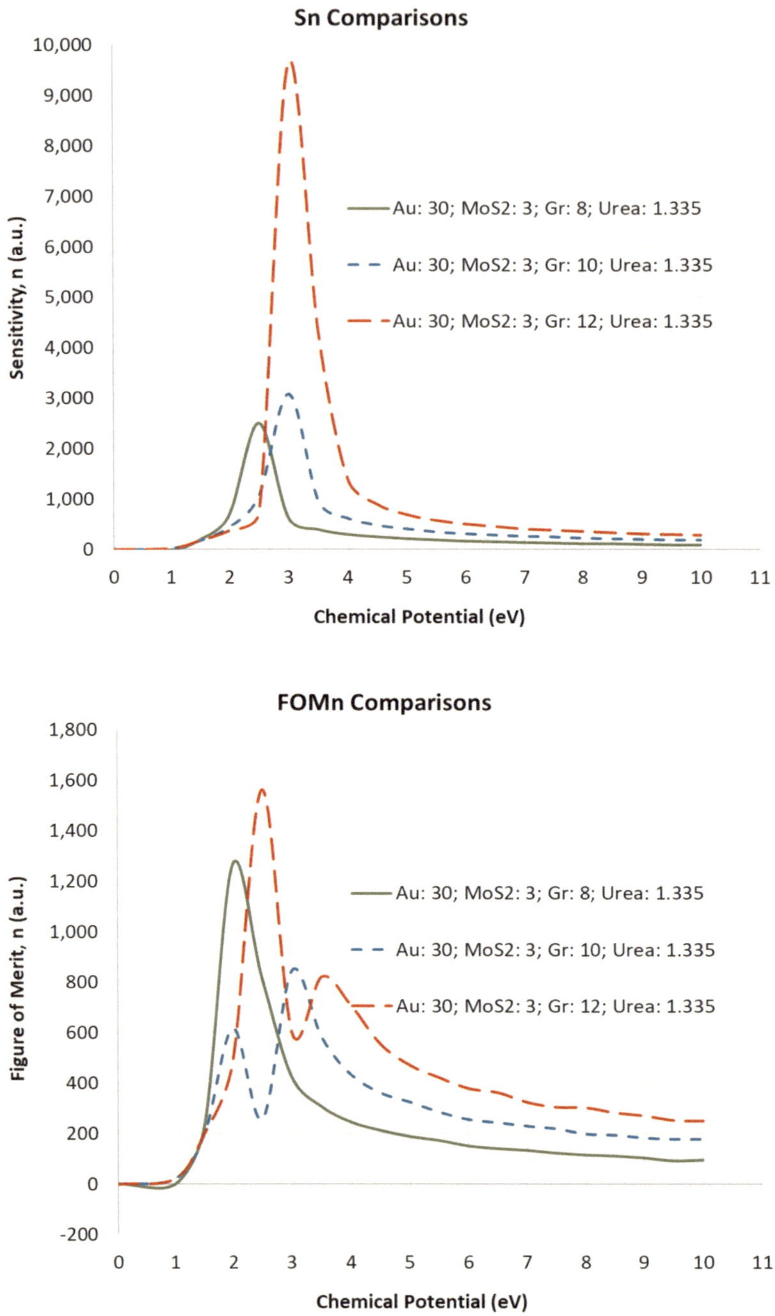


**Figure 8.**
*$S_n$ and $FOM_n$ curves for the sensor-sample: MgF2 ($n_p = 1.377$), followed by Au layer (30 nm thickness, $n_{Au}$ = 0.1726 + 3.4218 i), MoS2 (1 layer, $n_{MoS2} = 4.8046 + 0.84395\,i$, [18]), Gr (10 layers), and the sample considered is urea (10 nm thickness, $n_{urea} = 1.335$).*

| $n_{Urea}$→ | 1.332 | | | 1.335 | | | | | |
|---|---|---|---|---|---|---|---|---|---|
| **Gr layers →** | **8** | **10** | **12** | **8** | **8** | **10** | **10** | **12** | **12** |
| $\mu_c$ **(eV) ↓** | $\theta_{spr}$ | $\theta_{spr}$ | $\theta_{spr}$ | **Sn** | **FOM** | **Sn** | **FOM** | **Sn** | **FOM** |
| 0.001 | 49.50 | 49.46 | 49.46 | — | — | — | — | — | — |
| 1.000 | 49.50 | 49.50 | 49.54 | 0 | 0.00 | 24 | 22.22 | 24 | 22.99 |
| 1.500 | 48.89 | 48.92 | 48.92 | −204 | −226.67 | −180 | −200.00 | −180 | −200.00 |
| 2.000 | 47.45 | 48.13 | 48.35 | −684 | −1266.67 | −444 | −616.67 | −372 | −492.06 |
| 2.500 | 56.95 | 46.26 | 47.27 | 2496 | 815.69 | −1068 | −255.75 | −732 | −1564.10 |
| 3.000 | 51.34 | 58.64 | 20.59 | 624 | 422.76 | 3072 | 844.88 | −9636 | −600.15 |
| 3.500 | 50.65 | 52.34 | 62.39 | 396 | 305.56 | 960 | 579.71 | 4320 | 821.92 |
| 4.000 | 50.40 | 51.34 | 53.64 | 300 | 245.10 | 624 | 433.33 | 1404 | 709.09 |
| 4.500 | 50.22 | 50.90 | 52.16 | 252 | 212.12 | 480 | 360.36 | 900 | 555.56 |
| 5.000 | 50.15 | 50.65 | 51.55 | 216 | 187.50 | 408 | 323.81 | 696 | 471.54 |
| 5.500 | 50.08 | 50.51 | 51.19 | 192 | 172.04 | 348 | 284.31 | 576 | 421.05 |
| 6.000 | 50.00 | 50.40 | 50.98 | 168 | 150.54 | 312 | 254.90 | 504 | 378.38 |
| 6.500 | 49.97 | 50.33 | 50.80 | 156 | 139.78 | 288 | 242.42 | 456 | 361.90 |
| 7.000 | 49.93 | 50.26 | 50.69 | 144 | 133.33 | 264 | 229.17 | 408 | 323.81 |
| 7.500 | 49.90 | 50.18 | 50.58 | 132 | 122.22 | 252 | 218.75 | 384 | 304.76 |
| 8.000 | 49.86 | 50.15 | 50.51 | 120 | 114.94 | 228 | 197.92 | 360 | 303.03 |
| 8.500 | 49.82 | 50.11 | 50.44 | 120 | 111.11 | 216 | 193.55 | 336 | 282.83 |
| 9.000 | 49.82 | 50.08 | 50.40 | 108 | 103.45 | 204 | 182.80 | 312 | 270.83 |
| 9.500 | 49.79 | 50.04 | 50.36 | 96 | 91.95 | 192 | 177.78 | 300 | 252.53 |
| 10.000 | 49.79 | 50.00 | 50.29 | 96 | 95.24 | 192 | 177.78 | 288 | 250.00 |

*Note that $\theta_{spr}$ values for n = 1.335 are provided in **Table 1** for Gr: 8, 10, and 12 layers and not repeated here. The best achievable values for SPR angle, Rmin and FWHM are highlighted in red color. The SPR angle, Rmin and FWHM values are highlighted in blue color as comparison with the values for the Au(50 nm) - Gr sensor system as described in [11].*

**Table 2.**
*Chemical potential (eV) versus sensitivity and FOM for the sensor structure Au (30 nm) - $MoS_2$ (3 layers) - Gr, with the urea as sample thickness 10 nm, n = 1.335 compared with same sample urea at 10 nm thickness and n = 1.332.*

that the SPR angle shift happens toward the lower angles at lower $\mu_c$ values, and for comparison at different $\mu_c$ only the magnitude values should be used. Referring to FOM value of 1564 at 2.5 eV for Gr = 12 layers is a better indicator of the sensor's capability to detect the urea sample than only using the $\Delta\theta_{SPR}$, which clearly has a larger SPR angle shift value at 3 eV. This clearly indicates that a smaller FWHM is needed for a better measurable and distinguishable SPR curve signal. We see this same trend for Gr = 8 layers with FOM = 1267 and Gr = 10 layers with FOM = 845, where the FOM is a better indicator of the SPR signal detection. We see that at 8 eV, the FOM

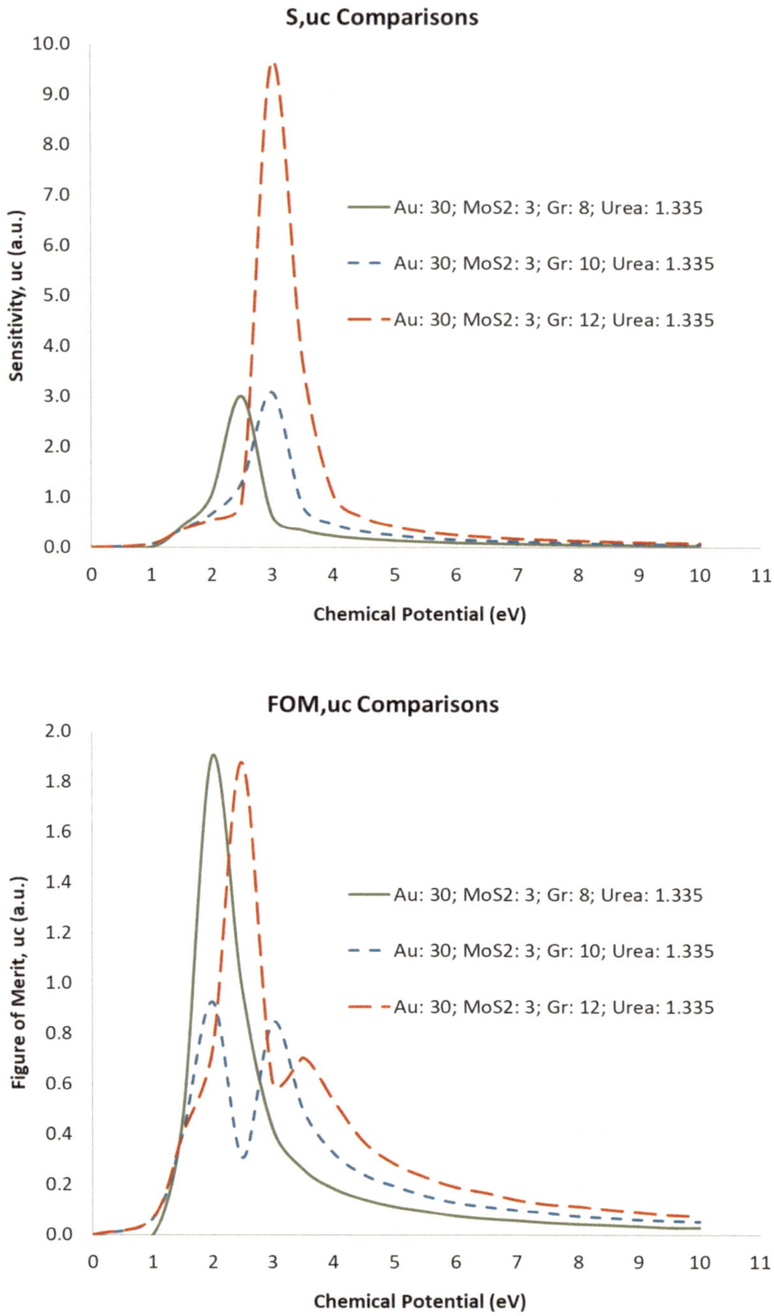


**Figure 9.**
*$S_{uc}$ and $FOM_{uc}$ curves for the sensor-sample: MgF2 ($n_p = 1.377$), followed by Au layer (30 nm thickness, $n_{Au}$ = 0.1726 + 3.4218 i), MoS2 (1 layer, $n_{MoS2} = 4.8046 + 0.84395\ i$, [18]), Gr (10 layers), and the sample considered is urea (10 nm thickness, $n_{urea} = 1.335$).*

values are rather in an upward trend from 115, 197 to 303 for Gr = 8, 10 &12 layers, which also indicates that the more number of graphene layers and the better sensitivity of the SPR system. **Table 3** shows the values of $S_{\mu c}$ and $FOM_{\mu c}$ for Gr = 8, 10 &12

| $n_{urea}$ → | 1.335 | | | 1.335 | | | 1.335 | | |
|---|---|---|---|---|---|---|---|---|---|
| **Gr layers →** | **8** | | | **10** | | | **12** | | |
| $\mu_c$ (eV) ↓ | $\theta_{spr}$ | $S\mu_c$ | FOM$\mu_c$ | $\theta_{spr}$ | $S\mu_c$ | FOM$\mu_c$ | $\theta_{spr}$ | $S\mu_c$ | FOM$\mu_c$ |
| 0.001 | 49.50 | — | — | 49.46 | — | — | 49.46 | — | — |
| 1.000 | 49.50 | 0.000 | 0.000 | 49.54 | 0.072 | 0.067 | 49.54 | 0.072 | 0.069 |
| 1.500 | 48.89 | −0.408 | −0.454 | 48.92 | −0.360 | −0.400 | 48.92 | −0.360 | −0.400 |
| 2.000 | 47.45 | −1.027 | −1.901 | 48.13 | −0.666 | −0.925 | 48.35 | −0.558 | −0.738 |
| 2.500 | 56.99 | 2.996 | 0.979 | 46.26 | −1.282 | −0.307 | 47.27 | −0.879 | −1.878 |
| 3.000 | 51.37 | 0.624 | 0.423 | 58.68 | 3.073 | 0.845 | 20.56 | −9.639 | −0.600 |
| 3.500 | 50.69 | 0.340 | 0.262 | 52.34 | 0.823 | 0.497 | 62.42 | 3.704 | 0.705 |
| 4.000 | 50.40 | 0.225 | 0.184 | 51.34 | 0.468 | 0.325 | 53.68 | 1.053 | 0.532 |
| 4.500 | 50.26 | 0.168 | 0.141 | 50.90 | 0.320 | 0.240 | 52.16 | 0.600 | 0.370 |
| 5.000 | 50.15 | 0.130 | 0.113 | 50.69 | 0.245 | 0.194 | 51.55 | 0.418 | 0.283 |
| 5.500 | 50.08 | 0.105 | 0.094 | 50.51 | 0.190 | 0.155 | 51.19 | 0.314 | 0.230 |
| 6.000 | 50.00 | 0.084 | 0.075 | 50.40 | 0.156 | 0.127 | 50.98 | 0.252 | 0.189 |
| 6.500 | 49.97 | 0.072 | 0.065 | 50.33 | 0.133 | 0.112 | 50.83 | 0.210 | 0.167 |
| 7.000 | 49.93 | 0.062 | 0.057 | 50.26 | 0.113 | 0.098 | 50.69 | 0.175 | 0.139 |
| 7.500 | 49.90 | 0.053 | 0.049 | 50.22 | 0.101 | 0.088 | 50.62 | 0.154 | 0.122 |
| 8.000 | 49.86 | 0.045 | 0.043 | 50.15 | 0.086 | 0.074 | 50.54 | 0.135 | 0.114 |
| 8.500 | 49.86 | 0.042 | 0.039 | 50.11 | 0.076 | 0.068 | 50.47 | 0.119 | 0.100 |
| 9.000 | 49.82 | 0.036 | 0.034 | 50.08 | 0.068 | 0.061 | 50.40 | 0.104 | 0.090 |
| 9.500 | 49.79 | 0.030 | 0.029 | 50.04 | 0.061 | 0.056 | 50.36 | 0.095 | 0.080 |
| 10.000 | 49.79 | 0.029 | 0.029 | 50.04 | 0.058 | 0.053 | 50.33 | 0.086 | 0.075 |

*The best achievable values for SPR angle, Rmin and FWHM are highlighted in red color. The SPR angle, Rmin and FWHM values are highlighted in blue color as comparison with the values for the Au(50 nm) - Gr sensor system as described in [11].*

**Table 3.**
*$S\mu_c$ (sensitivity wrt $\mu_c$), and FOM$\mu_c$ (figure of merit wrt $\mu_c$) values with reference to chemical potential for the sensor structure: Au (30 nm) - MoS2 (3 layers) - Gr, with the urea as sample thickness 10 nm, n = 1.335.*

layers and with reference to $\mu_c = 1$ meV. Highlighted in red color are the maximum values obtained. Here again, we see that $FOM_{\mu c}$ is a better indicator of the measurability of the SPR effect in a graphene-based SPR sensor, but with an added advantage of not having to change the sensor structure to achieve ultra-sensitivity to measure changes in the sample concentration upto 3/1000.

## 4. Conclusion

Graphene has been called a "wonder material" based on its electrical, mechanical, and chemical properties, and the numerous fields and applications have come to be used. It is not a surprise that the 2010 Nobel Prize for Physics was conferred to Drs. Andrew Geim and Konstantin Novoselov for their experimental work on graphene.

Based on its properties, graphene has become a much-preferred material to increase the sensitivity of SPR sensors. In this work, we present the idea of biasing a multi-material-graphene sensor surface structure to tune the refractive index of graphene thereby providing a significant enhancement in sensitivity. In this work, we have shown that the sensitivity of a Au-MoS2-Gr sensor system can be used to detect urea of 10 nm thickness sample layer and while the refractive index of the urea changes from n = 1.332 to 1.335, a change of 3/1000. Application of electrical bias across the metal-Gr layer structure tends to provide a strong excitation of the surface plasmon, which results in a very small reflectance at the SPR angle, which facilitates far better measurement. It should be noted that this increase in sensitivity is in addition to whatever the sensitivity original unbiased graphene-metal-based sensor structure may provide. Thus, the fact that the superior and more favorable electrical and electro-magnetic properties of graphene facilitate the application of electrical bias across it, resulting in a high sensitivity of the graphene-based SPR biosensor system.

## Acknowledgements

The authors thank the University of North Alabama Foundation, the College of Arts, Sciences, and Engineering, and the Department of Engineering and Industrial Professions at the University of North Alabama, Florence, AL for their financial support to pursue this research and publish this book chapter.

## Conflict of interest

The authors declare no conflict of interest.

## Abbreviations

| | |
|---|---|
| SPR | Surface Plasmon Resonance |
| Gr | Graphene |
| Au | Gold |
| Ag | Silver |
| MoS2 | Molybdenum disulfide |
| SiO2 | Silicon Dioxide |
| FWHM | Full Width at Half Maximum |
| FOM | Figure of Merit |
| Sn | Sensitivity wrt n - refractive index |
| $S\mu_c$ | Sensitivity wrt $\mu_c$ (chemical potential (eV)) |
| FOM $\mu_c$ | Figure of Merit wrt $\mu_c$ (chemical potential (eV)) |

## Appendix

The data of the SPR curves, SPR Angle, FWHM, and the Rmin are available at: SPR Data on GitHubhttps://github.com/RPGUNA/InTechOpen.git. More data is available upon request.

## Author details

Ravi Paul Gollapalli*, Tingyi Wei† and Jeremy Reid†
Department of Engineering and Technology, University of North Alabama, Florence, AL, USA

*Address all correspondence to: rgollapalli@una.edu

† These authors contributed equally.

## References

[1] Yamamoto M. Surface plasmon resonance (SPR) theory: Tutorial. Review of Polarography. 2002;**48**(3): 209-237. Available from: http://joi.jlc.jst.go.jp/JST.Journalarchive/revpolarography1955/48.209?from=CrossRef

[2] Homola J. Present and future of surface plasmon resonance biosensors. Analytical and Bioanalytical Chemistry. 2003;**377**(3):528-539. Available from: http://link.springer.com/10.1007/s00216-003-2101-0

[3] Piliarik M, Homola J. Surface plasmon resonance (SPR) sensors: Approaching their limits? Optics Express. 2009; **17**(19):16505. Available from: https://www.osapublishing.org/oe/abstract.cfm?uri=oe-17-19-16505

[4] Zhang H, Song D, Gao S, Zhang J, Zhang H, Sun Y. Novel SPR biosensors based on metal nanoparticles decorated with graphene for immunoassay. Sensors and Actuators B: Chemical. 2013;**188**: 548-554. Available from: http://www.sciencedirect.com/science/article/pii/S0925400513008526

[5] Lee KL, Lee CW, Wang WS, Wei PK. Sensitive biosensor array using surface plasmon resonance on metallic nanoslits. Journal of Biomedical Optics. 2007; **12**:1-5. DOI: 10.1117/1.2772296

[6] Špringer T, Ermini ML, Špačková B, Jabloňků J, Homola J. Enhancing sensitivity of surface plasmon resonance biosensors by functionalized gold nanoparticles: Size matters. Analytical Chemistry. 2014;**86**:10350-10356. DOI: 10.1021/ac502637u

[7] Banerjee AN. Graphene and its derivatives as biomedical materials: Future prospects and challenges. Interface Focus. Jun 2018;**8**(3): 20170056. Available from: https://royalsocietypublishing.org/doi/10.1098/rsfs.2017.0056

[8] Nurrohman DT, Wang YH, Chiu NF. Exploring graphene and MoS2 chips based surface plasmon resonance biosensors for diagnostic applications. Frontiers in Chemistry. 2020;**8**:00728. Available from: https://www.frontiersin.org/article/10.3389/fchem.2020.00728

[9] Wu L, Chu HS, Koh WS, Li EP. Highly sensitive graphene biosensors based on surface plasmon resonance. Optics Express. 2010;**18**(14):14395. Available from: https://www.osapublishing.org/oe/abstract.cfm?uri=oe-18-14-14395

[10] Choi SH, Kim YL, Byun KM. Graphene-on-silver substrates for sensitive surface plasmon resonance imaging biosensors. Optics Express. 2011;**19**(2):458-466. Available from: http://www.opticsexpress.org/abstract.cfm?URI=oe-19-2-458

[11] Gollapalli RP. Enhanced sensitivity in graphene-based SPR biosensors using electrical bias. Optics Letters. 2020;**45**(10):2862. Available from: https://www.osapublishing.org/abstract.cfm?URI=ol-45-10-2862

[12] Wei T, Reid JL, Gollapalli RP, Vickery D, Thompson DB, Gren CK. Dynamic tunability of sensitivity of graphene based surface plasmon resonance sensor. In: OSA Optical Sensors and Sensing Congress 2021 (AIS, FTS, HISE, SENSORS, ES). Optica Publishing Group; 2021. p. STu4F.3. Available from: http://opg.optica.org/abstract.cfm?URI=Sensors-2021-STu4F.3

[13] Hanson GW. Dyadic Green's functions and guided surface waves for a

surface conductivity model of graphene. Journal of Applied Physics. 2008;**103**(6): 064302. Available from: http://aip.scitation.org/doi/10.1063/1.2891452

[14] Novoselov KS, Geim AK, Morozov SV, Jiang D, Zhang Y, Dubonos SV, et al. Electric field effect in atomically thin carbon films. Science. 2004;**306**(5696): 666. Available from: http://science.sciencemag.org/content/306/5696/666.abstract

[15] Matthaiakakis N, Mizuta H, Charlton MDB. Strong modulation of plasmons in graphene with the use of an inverted pyramid array diffraction grating. Scientific Reports. Jul 2016;**6**(1):27550. Available from: http://www.nature.com/articles/srep27550

[16] Stauber T, Peres NMR, Geim AK. Optical conductivity of graphene in the visible region of the spectrum. Physical Review B. Aug 2008;**78**(8):085432. Available from: https://link.aps.org/doi/10.1103/PhysRevB.78.085432

[17] Hu H, Zavabeti A, Quan H, Zhu W, Wei H, Chen D, et al. Recent advances in two-dimensional transition metal dichalcogenides for biological sensing. Biosensors & Bioelectronics. 2019;**142**: 111573. Available from: https://www.sciencedirect.com/science/article/pii/S0956566319306529

[18] Song B, Gu H, Fang M, Chen X, Jiang H, Wang R, et al. Layer-dependent dielectric function of wafer-scale 2D MoS. Advanced Optical Materials. 2019; **7**(2):1801250. Available from: https://onlinelibrary.wiley.com/doi/10.1002/adom.201801250

[19] Xu H, Wu L, Dai X, Gao Y, Xiang Y. An ultra-high sensitivity surface plasmon resonance sensor based on graphene-aluminum-graphene sandwich-like structure. Journal of Applied Physics. 2016;**120**(5):053101. Available from: http://aip.scitation.org/doi/10.1063/1.4959982

Chapter 10

# Graphene Oxide-Based Membranes as Water Separation: Materials, Preparation, Characteristics, and Applications

*Aster Woldu*

## Abstract

Membrane-based separation technology has attracted great interest in many separation fields due to its advantages of easy-operation, energy-efficiency, and easy scale-up and environmental friendliness. The development of novel membrane materials and membrane structures is an urgent demand to promote membrane-based separation technology. Graphene oxide (GO), as an emerging star nano-building material, has showed great potential in the membrane-based separation field. In this review paper, it is briefly reviewed the preparation and characterization of GO. Then, the preparation method, characterization, and type of GO-based membrane are summarized. Before that the general concept behind membrane technology is presented.

**Keywords:** graphene oxide (GO), graphene, graphene-oxide membrane

## 1. Introduction

Unquestionably; water is the source of life and one of the most important material resources for human existence and advancement. Although 71% of the earth's surface is covered with water; approximately 98% of our water is salty and only 2% is fresh. Of that 2%, almost 70% is snow and ice, 30% is groundwater and less than 0.5% is surface these are freshwater resources that can be directly used by humans, such as river water, freshwater lakes, and shallow groundwater [1]. Meaning; there is no surplus fresh water in the planet. In addition to this, modernization that demands the fast development of industries and increasing human activities, forced many harmful inorganic and organic pollutants to be released into water, which extremely jeopardizes the available freshwater resource and ecological environment. Today according to the world population clock, the population exceeds 7 billion and will reach 10 billion by 2050. These all indicates pure drinking water would be a major problem all over the world, especially for the developing countries [1].

As a kind of technology which can increase the amount of freshwater, desalination of the salty water and treating the polluted water have become a strategic choice to solve the crisis of water resources. Currently, the global volume of desalinated water



IntechOpen

has exceeded 90 million $m^3$/day, alleviating water shortages which have affected over 200 million people [1, 2]. Reusable water obtained from treating the wastewater is not as such considerable [3, 4].

To desalinate water a number of process technologies are used. The mentioned techniques are (1) reverse osmosis (RO), distillation, freezing, hydration and solvent extraction, etc. or (2) ion exchange, electro-dialysis (ED), adsorption capacitance, pressure impregnation and forward osmosis (FO) technology [1–3]. Out of the mentioned technologies; three of the processes require the use of semipermeable membranes.

Literature indicated membrane technology has become a dignified separation technology over the past few decades [2] and from **Figure 1** the polymer based membrane technology is in the forefront of water purification and desalination, owing to its advantages of low energy consumption (as compared to other technology, low investment cost, ease of operation and possibility for continuous operation and inherent simplicity) [3, 5], but is plagued with some bottlenecks like most of them tend to foul, have low resistance to chlorine, strong acids/alkaline, high temperature and organic solvents, and suffer from aperture shrinkage under high pressure [1, 3]. And despite large scale seawater desalination plants have already confirmed their much needed success, the widespread implementation of these plants is held back due to their high energy costs. One approach to resolve the mentioned problems is through enhancing a membrane used in separation process. A tremendous amount of effort has been paid to develop new membranes and develop novel membrane structures with greater chemical stability, thermal stability, water permeability, as well as high selectivity, which in turn yield less energy consumption [6–10].

Since 2004; researcher focused on carbon-based materials especially one material-graphene. Graphene and its derivative graphene oxide (GO) and carbon nanotubes

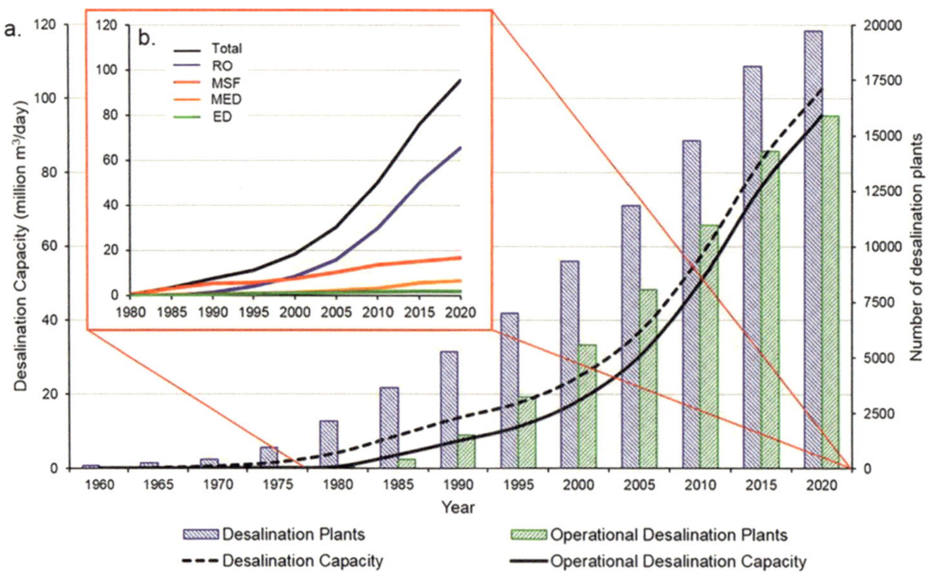


**Figure 1.**
*Trends in global desalination by (a) number and capacity of total and operational desalination facilities and (b) operational capacity by desalination technology (https://www.desaldata.com/).*

(CNTs) have demonstrated notable potential in the field of membrane. The basic characters for their suitability are their strong mechanical strength, high resistance to strong acids/alkaline and organic solvents, and easy availability. Among them, GO selected for evolving nano-building materials for the manufacturing of novel separation membrane due to its high mechanical strength, high chemical inertness, nearly frictionless surface, its flexibility, suitability for large-scale production and its cost-effectiveness [5–10].

In this mini review first an introduction about membrane, graphene and graphene oxide (GO) are provided. Then it is tried shortly to discusses points on the preparation and characterization of GO. Finally the preparation method, characterization, and type of GO-based membrane are discussed.

## 2. Briefing on membrane separation

Membrane separation is a technology in which selectively materials are screened out via pores or minute gaps presented in the molecular arrangement of membrane structure. Membrane separations are basically classified by pore size and by the separation driving force. These classifications are: microfiltration (MF), ultrafiltration (UF), ion-exchange (IE), and reverse osmosis (RO), and forward osmosis (FO) [11].

Membrane separations are primarily used to separate solids (either particulate or in solution) from liquid. The technology also used in effect liquid-liquid separation and gas-to-liquid infusion processes. Reverse osmosis (RO), ultra-filters (UF), and micro-filters (MF) are membrane technologies that employ pressure across the membrane as the driving force for separation (a trans-membrane pressure or TMP). Basically UF practical in concentrating and purifying virus, bacteria, fermentation broths, colloids, and particles, as well as dissolved high molecular weight polymer. Since membrane separations can also be accomplished at below ambient temperatures; they are suitable in the manufacture of degradable materials such as pharmaceuticals and foods. MF is engaged for separation and purification of suspended solids, colloidal particles, and bacteria from liquids and it is also used in the screening of particles and bacteria from gases. Non-filtration membrane technologies are dialysis and electro dialysis. Dialysis uses concentration differences as the separating driving force and the electro dialysis uses electric potential differences as the separating driving force for the separation. An artificial kidney is a specific example of dialysis where blood impurities are effectively removed for renal failure patients.

## 3. Contribution of membrane technology for water treatment

Reverse osmosis (RO), forward osmosis (FO), nanofiltration (NF), ultrafiltration (UF), microfiltration (MF), and particle filtration are working in water treatment process to treat the raw or wastewater depends on their pore sizes. **Figure 2** summarizes the various membrane filtration processes related to common materials that would be filtered out through each process during water treatment.

**Figure 3** illustrates the application of FO and reverse osmosis process for water purification of sea water and wastewater.

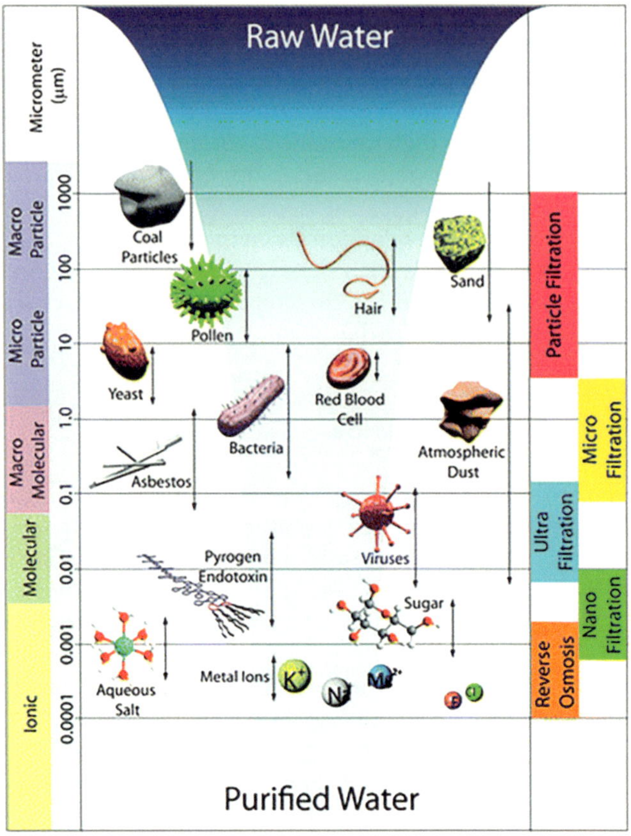


**Figure 2.**
*Illustration of different membrane filtration spectrum.*

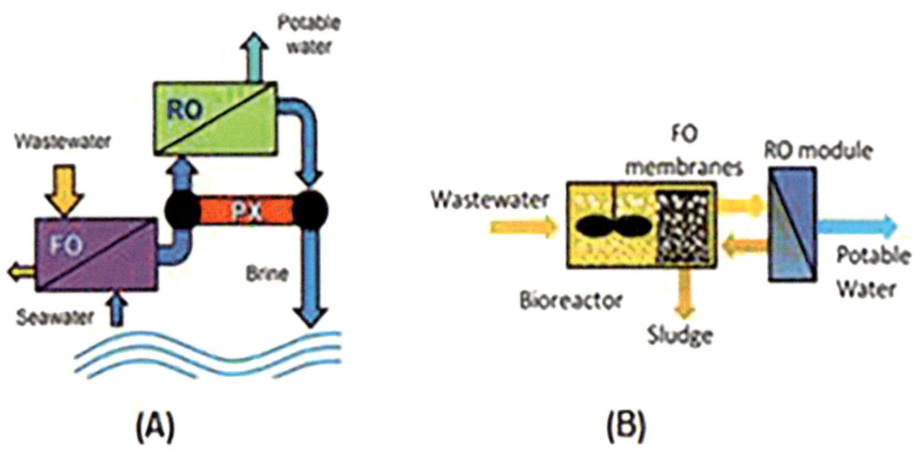


**Figure 3.**
*Examples of application of membrane technology for water treatment. (A) Diagram of an FO-RO system for integrated SWRO and wastewater recovery. Water from the wastewater is extracted by osmosis through the FO process and dilutes the seawater before entering the RO process. (B) OMBR (osmotic membrane biological reactor) system comprising a bioreactor containing submerged FO membranes and an RO unit that re-concentrates the draw solution and produces water.*

## 4. Conventional membrane material

Almost all industrial membrane processes are carried by organic membranes made up of natural (wool, rubber (polyisoprene)) or synthetic (polyamide, polystyrene and poly-tetra-fluoro-ethylene (Teflon) polymers. (http://www.separationprocesses.com; [6]). Membranes can also be manufactured from other non-polymeric materials. Such membranes materials can be inorganic membranes (like metal, ceramic, carbon, and zeolites) and liquid membranes. But their application in water purification is not exhibited (http://www.separationprocesses.com).

For purification of water; currently the polymeric membrane governed the membrane market including real-world application and academic research due to its advantages of energy-efficiency, easy-operation, low-cost, and inherent simplicity. However, there are limitations related to the conventional membranes exhibited in most practical applications. Most of them tend to foul have low resistance to chlorine, strong acids/alkaline, high temperature and organic solvents, high contaminant permeation relative to stringent selectivity requirement and suffer from aperture shrinkage under high pressure [6]. The strong trade-off relation between membrane selectivity and permeability is a common challenge for all of polymeric membranes.

Because of the mentioned limitations the development of novel membrane materials is a major research thrust for academia, industry, and national laboratories. The desirable membrane properties, which are constantly searched by the researchers are good mechanical strength, superior chemical stability, thermal stability, water permeability, as well as high selectivity [6].

## 5. Graphene and graphene-oxide

### 5.1 Graphene

Since 2004, the Nobel Prize winning material, graphene is taken as one of the most wonderful achievements in the field of science and technology [7, 12]. Graphene is an atomically-thin (0.35 nm in thickness), it is a two-dimensional sheet with a honeycomb structure made up of $sp^2$ hybridization carbon atoms which are linked together with strong sigma keys [7]. Graphene has unique properties that make it a core of scientific research that many of them can be transformed into practical [5, 7–9]. These valuable properties includes (1) a highly specific surface area [5, 7], as reported by Zhu et al. [13]; specific surface area (SSA) values of carbon materials obtained from GO have been well below 2630 $m^2/g$, but the specific surface area of common active carbon is only 1500 $m^2/g$ [12].

It has also a remarkable elasticity and mechanical strength; atomic force microscopy (AFM) measures the performance of freestanding monolayer graphene membrane based on Nano indentations; the result showed it has a breaking strength of 42 $N{\cdot}m^{-1}$ and a Young's modulus of TPa 1.0. even if graphene is an extremely strong material, still an external mechanical load can change the electronic properties of graphene thus, it is possible to affect its field emission performance. Graphene's capability to absorb pressure can also be affected by different degrees of axial compression. The varying sizes of buckling stress and strain are measured using a cantilever beam. When graphene is used as a membrane material, it can provide a stronger support force and adjustable sheet spacing. Therefore, the mechanical strength and controllability of the membrane can be improved. Graphene also exhibited excellent molecular barrier abilities and superior thermal and electrical conductivity [8–11]. Frequently, graphene and its derivatives are used in super capacitor, fuel cells, capacitive deionization, desalination, and others [13]. Different literature also reported graphene derivatives which can be integrated with other materials such as inorganic nanostructures, organic crystal, polymers, organic framework, biological materials, and carbon nanotubes to improve specific properties of the materials (Yi [11]).

### 5.2 Graphene-oxide (GO)

The structure of GO similar to graphene but it also contains hydroxyl (–OH), alkoxy (C–O–C), carbonyl (C=O), carboxylic acid (–COOH), and other oxygen-based functional groups. GO has a non-stoichiometric general formula of the type $C_xH_yO_z$. See **Figure 4** it is a suitable nanoparticle to improve the hydrophobicity of the membrane. The functional groups making it more dispersed in the polymeric solution. That means, if it is incorporated in the membranes, it can be improve properties for water purification. GO-incorporated membranes consist of a high mechanical strength and thermal stability [11]. The GO-incorporated membrane can enhance water transport even in low-pressure applications [14].

The most important property of GO nano-sheets is antifouling during operation due to negative charge and high hydrophobicity [7, 11]. It is believed that the GO-incorporated membranes have improved fouling resistance by reducing surface roughness and increasing hydrophilicity [7]. The GO-incorporated membrane shows high water permeability in various applications such as NF, RO, FO, and PRO processes [11]. Although the benefits of GO has been reported many, the material is still

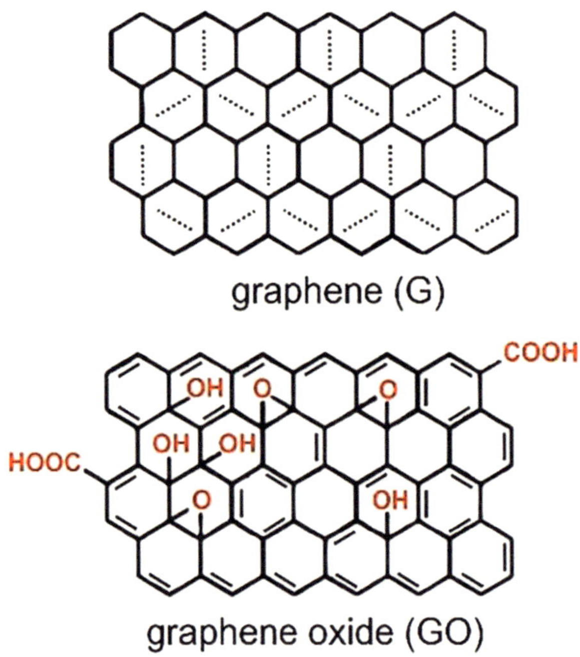


**Figure 4.**
*Depicted generic chemical and physical structures of graphene-based materials.*

expensive and is relatively difficult to manufacture at a larger capacity. Therefore, the manufacture of the material expected to well develop to decrease the cost and complication of manufacturing [11, 12].

## 6. Preparation of graphene oxide

The preparation process of GO essentially comprise two steps: oxidation of graphite and exfoliation of graphite oxide. In literature different methods are reported for the preparation of GO [12]. **Table 1** summarized different methods used for the preparation of GO.

| Year | Methods' name | Oxidant | Acid used | Reaction time | Note: limitation and advantage | Reference |
|---|---|---|---|---|---|---|
| 1859 | Brodie | $KClO_3$ | $HNO_3$ | 3–4 days | Time consuming and generated toxic gas ($ClO_2$), which was unsafe and harmful to the environment. | Smith et al. [12] |
| 1898 | Staudenmaier | $KClO_3$ | $H_2SO_4$, $HNO_3$ | 1–10 days | Improved Brodie's method by adding $KClO_3$ in multiple aliquots during the oxidation course and further acidifying the mixture by adding concentrated sulfuric acid ($H_2SO_4$)<br>More practical and convenient for the production of GO with comparable oxidation degree<br>Produced toxic gases ($ClO_2$, $NO_x$) and is not environmentally friendly. | Smith et al. [12] |
| 1937 | Hofmann | $KClO_3$ | $H_2SO_4$, $HNO_3$ | 4 days | A modification of Brodie's method, it substituted fuming $HNO_3$ with non-fuming $HNO_3$ during the oxidation course<br>Produced toxic gases ($ClO_2$, $NO_x$) and is not environmentally friendly | Smith, et al. [12] |
| 1958 | Hummers and Offeman | KMnO4 | $NaNO_3$, $H_2SO_4$ | 2 h | More efficient and less time consuming<br>Widely used in current research<br>Toxic gas $NO_x$, $Mn^{2+}$ in GO | |

| Year | Methods' name | Oxidant | Acid used | Reaction time | Note: limitation and advantage | Reference |
|---|---|---|---|---|---|---|
| 1999 | Modified Hummers | KMnO4 | $K_2S_2O_8$, $P_2O_5$, $H_2SO_4$ | | This included two oxidation procedures. Preoxidized the graphite in a mixing solution of concentrated $H_2SO_4$, $K_2S_2O8$, and $P_2O_5$ at 80°C. Then the mixture was washed and dried at ambient temperature. After that, the mixture was ulteriorly oxidized by Hummers' method Controlling the temperature is major factor $NO_2$, $N_2O_4$ generated, hazardous heavy metal $Mn^{2+}$ introduced in the preparation process | |
| 2010 | Improved Hummers | KMnO4 | $H_2SO_4/H_3PO_4$ | 12 h | Hybrid of $H_2SO_4/H_3PO_4$ with volume ratio of 9:1 was used as the mixed acid Simpler and higher yielding, and generated no toxic gas making it possible for large-scale production of GO. Hazardous heavy metal Mn2+ introduced in the preparation process | |
| 2015 | Iron-based green method | $K_2FeO_4$ | H2SO4 | 1 h | Less time consuming (1 h) and enabled the recycle of $H_2SO_4$, which decreased the pollution to environment. $Fe^{3+}$ in GO | Peng et al. [15] |

**Table 1.**
*Methods for the preparation of GO.*

## 7. GO membrane fabrication

The final step in fabricating GO membranes involves integrating GO with the support. Different supports are used [11] for this purposes. After selecting the support, there are also various routes for this integration step. The vacuum filtration shown in **Figure 5** is the most applied method [11, 12]. The process involves filtering a suspension of GO in a solvent (usually water) through the support structure, with the retained GO sheets stacking up on each other and effectively being deposited on its surface and forming a thin film on the support. This layering step determines the GO membrane performance. Similar to vacuum filtration in principle is drop casting,

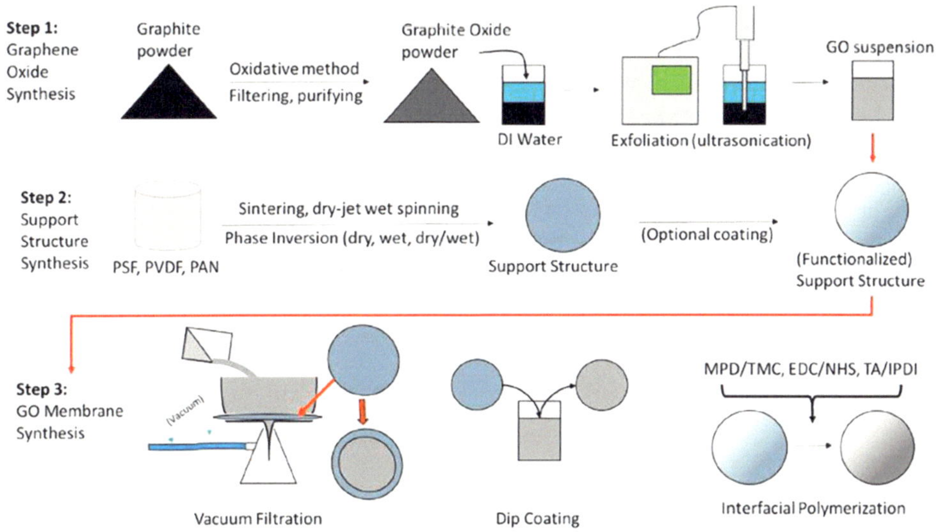


**Figure 5.**
*Steps applied to manufacture GO membrane [11].*

which deposits GO onto the support via droplets of concentrated solutions. Upon the evaporation of the solvent, a layer of GO is left behind [8, 9].

Another technique to form membranes from GO is dip coating, which involves dipping the support into a solution of GO where it is coated with a thin film [11]. The uses of spray-coating, spin-coating and chemical vapor deposition (CVD) methods are also reported to be applied for GO membrane manufacturing [11, 12].

## 8. Conclusion

Because of many advantage like easy-operation, energy-efficiency, and easy scale-up, and environmental friendliness; membrane-based separation technology attracted many separation fields. But the membrane fabricated afflicted with bottlenecks like- fouling, have low resistance to chlorine, strong acids/alkaline, high temperature and organic solvents, and suffer from aperture shrinkage under high pressure and high energy cost. One of the methods to resolve the mentioned problems is through enhancing a membrane castoff in separation process.

A tremendous amount of effort has been paid to develop new membranes and develop novel membrane structures with greater chemical stability, thermal stability, water permeability, as well as high selectivity, which in turn yield less energy consumption. Since 2004 researcher focused on carbon-based materials especially one material—graphene. Graphene, and its derivative graphene oxide (GO) and carbon nanotubes (CNTs). They have been demonstrated notable potential in the field of membrane. The basic characters for their suitability are: their strong mechanical strength, high resistance to strong acids/alkaline and organic solvents, and easy availability. Among them, GO selected for evolving nano-building materials for the manufacturing of novel separation membrane due to its high mechanical strength, high chemical inertness, nearly frictionless surface, its flexibility, suitability for large-scale production and its cost-effectiveness.

The Nobel Prize winning material, graphene is taken as one of the most wonderful achievements in the field of science and technology. It is an atomically-thin (0.35 nm in thickness), it is a two-dimensional sheet with a honeycomb structure made up of $sp^2$ hybridization carbon atoms which are linked together with strong sigma keys.

Preparation of GO from graphene essentially comprise two steps: oxidation of graphite and exfoliation of graphite oxide and is carried out by simply six different kinds of oxidation process. To get GO membranes it need only a final steps of integrating GO suspension with the support by vacuum filtration, dip coating or interfacial polymerization.

## Author details

Aster Woldu
Adama University, Adama, Ethiopia

*Address all correspondence to: aster.woldu@research.astu.edu.et

## References

[1] Jones E et al. The state of desalination and brine production: A global outlook. Science of the Total Environment. 2019;**657**:1343-1356

[2] Lattemann S, Höpner T. Environmental impact and impact assessment of seawater desalination. Desalination. 2008;**220**:1-15

[3] Elimelech M et al. The future of seawater desalination: Energy, technology, and the environment. Science. 2011;**333**(6043):712-717

[4] Parfit M, Graves W. Water: The power, promise, and turmoil of North America's fresh water. National Geographic Society. 1993;**4**:6-13

[5] Zhang N et al. Review on structural control and modification of graphene oxide-based membranes in water treatment: From separation performance to robust operation. Chinese Journal of Chemical Engineering. 2019;**27**:1348-1360

[6] Hu M, Mi B. Enabling graphene oxide nano-sheets as water separation membranes. Journal of Environmental Science and Technology. 2013;**47**:3715-3723

[7] Song N. et al. A review of graphene-based separation membrane: Materials, characteristics, preparation and applications. 2018.

[8] Joshi RK et al. Precise and ultrafast molecular sieving through graphene oxide membranes. Science Reports. 2014;**343**:752. DOI: 10.1126/science.1245711

[9] Nair RR et al. Unimpeded permeation of water through helium-leak–tight graphene-based membranes. Science Reports. 2012;**335**:442-443

[10] Hung W-S et al. Tuning the interlayer spacing of forward osmosis membranes based on ultrathin graphene oxide to achieve desired performance. Carbon. 2018

[11] Jiang Y et al. A review of recent developments in graphene enabled membranes for water treatment. Environmental Science: Water Research & Technology. 2016;**2**:915

[12] Smith AT et al. Synthesis, properties, and applications of graphene oxide/reduced graphene oxide and their nanocomposites. Nano Materials Science. 2019;**1**:31-47

[13] Zhu Y et al. Carbon based super capacitors produced by activation of graphene. Science. 2011;**332**:1537

[14] Yang Z et al. The preparation and study of ethylene glycol-modified graphene oxide membranes for water purification. Polymers. 2019;**11**:188. DOI: 10.3390/polym11020188

[15] Peng L et al. An iron-based green approach to 1-h production of single-layer graphene oxide. Nature Communications. 2015;**6**:5716

Chapter 11

# Graphene Composite Cutting Tool for Conventional Machining

*Gaurav Saini, Sarabjeet Singh Sidhu and Simranpreet Singh Gill*


## Abstract

Cutting is an important process in the manufacturing industry and cutting tool is an important element in machining. It is essential to use good quality cutting tools in arrange to maintain the quality of a product. To retain the performance of cutting tool, various techniques have been utilized like cutting fluid, cutting under MQL, coating, multilayer coating, cryoprocessing, different types of surface texturing, different types of solid lubricants, etc. All these processes have a great impact to enhance the mechanical, thermal, and tribological properties in case of conventional machining process. Nowadays composite engineered materials are very successful in metal cutting industry due to its wear-related application and excellent mechanical and thermal properties. A very few research has been carried out on graphene mixed composite tool material, which has very high demand in manufacturing industries, due to its application as a cutting tool material for machining of Al, copper, or high strength carbon steel. In the end, challenges in the processing of tungsten carbide graphene mixed self-lubricated tool have been identified from the literature. In parallel, the latest improvements to enhance the properties of tungsten carbide-cobalt cutting tool with graphene mixed are reviewed.


**Keywords:** tungsten carbide, tool life, powder metallurgy, spark plasma sintering, graphene nano-powder

## 1. Introduction

Today, almost every manufacturing industry is focused on improving productivity and reducing operational cost associated with machining operation [1]. In the current machining scenario, this can be achieved by implementing several changes such as providing required training to skilled workers, procuring better material, and using better quality cutting tools. Cutting tools being the heart of a machining system requires attention while selection of its material and other specifications such as tool life due to its high cost. The quality of the tools plays an important role in the machine-building and energy sectors, and many other industries. They need to be characterized by high productivity, wear resistance, and technological effectiveness.

Cutting tools are made of different types of materials such as carbon tool steel, high-speed steel, cemented carbide, boron nitride, and diamond tools [2]. All these materials have different mechanical and tribological properties, but particularly for steel cutting applications, cemented tungsten carbide is a commonly used material in the manufacturing industry. Various studies have been made over the generation for further improvement of cemented carbide properties to meet the severe cutting conditions. Various studies have also reviewed different ways to improve tool life by various types of coatings on tungsten carbide–based cutting tools [3]. However, the results of these studies vary due to different process parameters and manufacturing technology. In general, materials for cutting tools must possess high hardness, sufficient toughness, as well as hot strength in order to withstand high working temperature during the machining process. Also, hardness is one of the important factors that determine the life of tool at extreme temperature conditions. Therefore, high-speed steel and cast-cobalt alloys are no longer preferable as cutting tools [4].

Cemented carbides were initially introduced in the 1930s to overcome the challenge of high cutting speed that was impossible with the high-speed steel tool material. Cemented carbide shows high hardness (which is adequately stable over a wide range of temperature), high elastic modulus, high thermal conductivity and low thermal expansion. It is most widely used in machining, drilling and other related applications [5]. The use of tungsten carbide cutting tool is growing faster in metal cutting industry. Tungsten cemented carbide typically comprises tungsten carbide (WC) particles bonded together in cobalt (Co) matrix. WC-Co properties depend upon hard and brittle carbide, while cobalt as a metal binder provides ductility and toughness to the composite. Today, hard metal industry is focusing on the advancement of all cutting tools like high-speed steel, cemented carbide, and ceramic to obtain better grained composite to maximize hardness while maintaining reasonable toughness in order to meet the extreme cutting conditions. As compared to speed steel and ceramic, cemented carbide tool has high hardness, high wear resistance, good strength, and toughness and retains hardness at high temperature [6].

There are various types of techniques that have been utilized to improve the performance of cutting tool like cutting fluid, coating, multilayer coating, cryoprocessing, different types of surface texturing, and applied solid lubricants as shown in **Figure 1**. There are many studies that concern the enhancement of properties of various cutting tool, but there is no dedicated summarization of composite cutting tool works. Hence, the current study summarizes the different work done to improve the properties of WC-Co composite utilizing different consolidation and sintering techniques.

## 2. WC-Co-based cutting tool materials

WC-CO has become one of the most common materials for many applications requiring high temperature and high wear resistance like cutting tool in manufacturing industry [7]. It also includes machining of cast iron, various non-ferrous metals, etc. WC-based composite tool materials normally possess higher hardness, strength, thermal conductivity, chipping resistance, and plastic deformation resistance and have been universally used in high-speed cutting. These attractive properties gave it great potential to be used as high-speed cutting tool materials. But the application of WC as cutting tools is severely limited due to low fracture toughness. Cemented

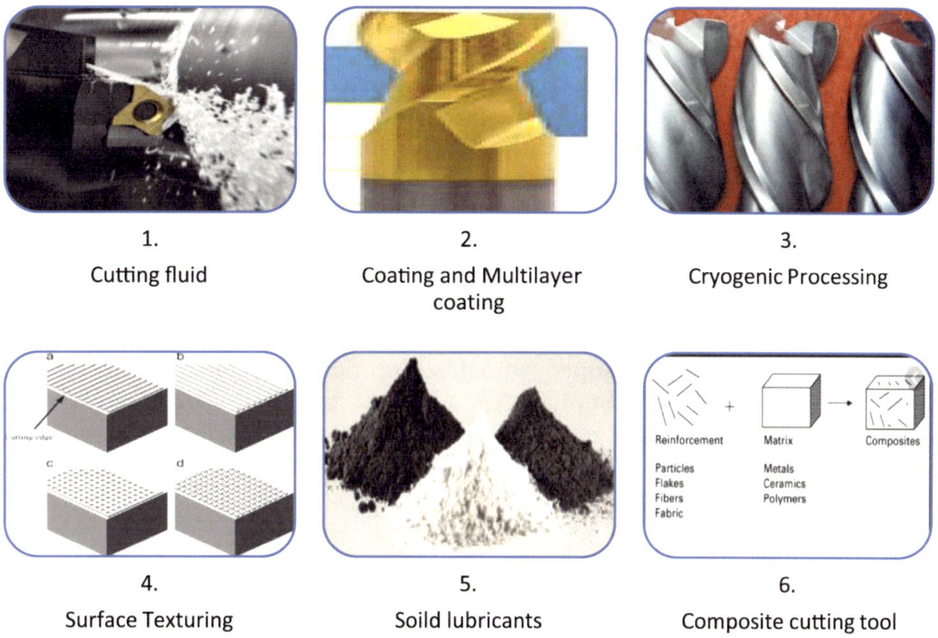


**Figure 1.**
*Various techniques to improve the performance of cutting tool.*

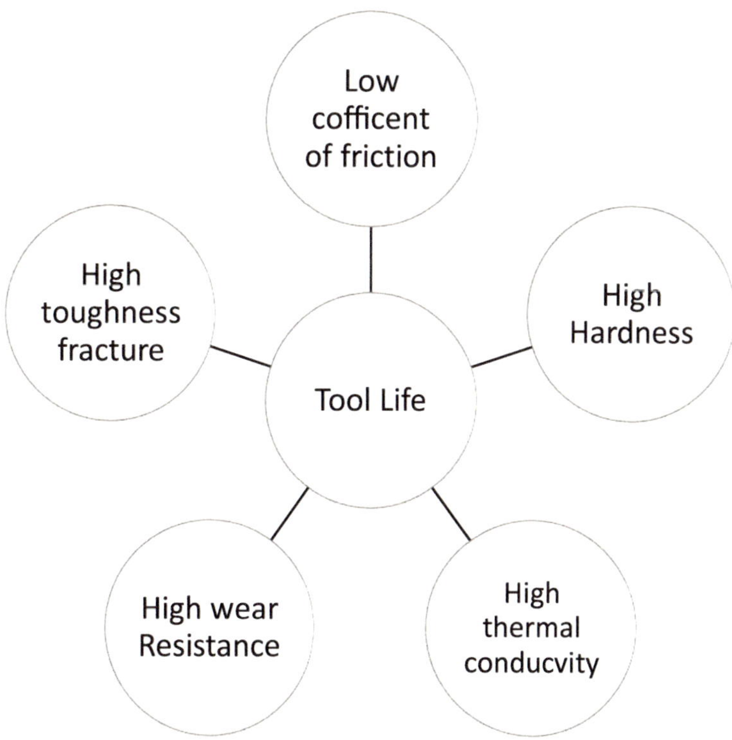


**Figure 2.**
*Tool life depends upon various factors.*

carbides belong to the most common and the longest-used tool materials produced by powder metallurgy methods [8, 9]. Sintered carbides are characterized by their high strength and abrasion resistance and include one or more high-melting metal carbides constituting the basic component together with the metallic binding phase. The basic component of cemented carbides is WC, which, depending on the manufacturer and group of material applications, can constitute from 50% to 90% by weight of the sintered content. Cemented carbides have replaced high-speed steels in metalworking and mining. Cemented carbides are metal-ceramic composites that consist of hard tungsten, titanium, and tantalum carbide grains located in a ductile matrix (binder) based on cobalt or nickel and have a unique combination of high hardness, wear-resistance, and toughness [10]. By now, a large number of cemented carbide grades have been developed, with diverse combinations of components both in the carbide phase and in the binder. Nowadays market distribution of cutting tool material has shown that cemented carbide has a dispersion of 53%, high-speed steel has a dispersion of 20% [11]. As we know, tool life of cutting tool depends upon various factors as shown in **Figure 2**.

## 3. Classification of composite cutting tool

Composite is the combination of two main constitutions namely matrix and reinforcement with the specific end goal to enhance the properties [12]. In recent years, cutting tool industry move from metal cutting tool to composite cutting tool. The function of the matrix is to hold the reinforcement particles in position by surrounding and supporting them. The reinforcements generally have an impact on mechanical and physical properties or any other tailored property enhanced from the matrix material [13]. The combination of a wide variety of reinforcement materials for specific effect on matrix allows the researchers to choose an optimum combination of materials to make a product and also classification of composite material as shown in **Figure 3**.

### 3.1 Classification of composites based on matrix material

#### *3.1.1 Metal matrix composite (MMC)*

A metal matrix composite is a combination of two or more constituents, of which at least one should be a metal, and the other may be a metal, ceramic or organic

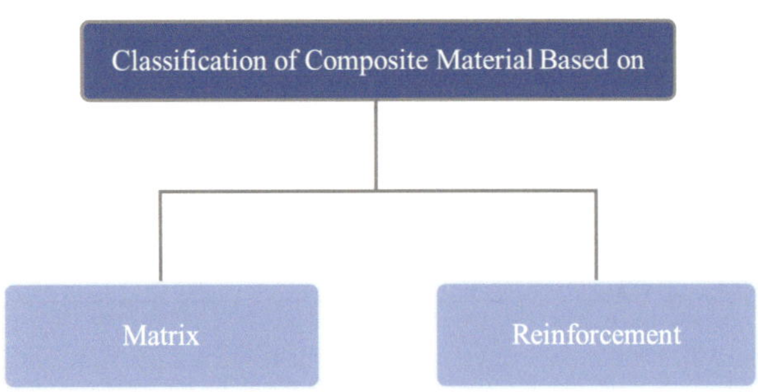


**Figure 3.**
*Classifications of composite materials.*

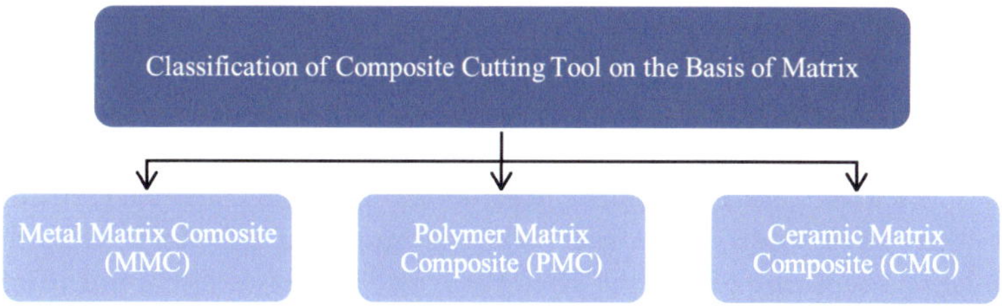


**Figure 4.**
*Classification of composite cutting tool.*

compound [14]. Various matrix used in MMC are aluminum, copper, magnesium, and iron. Oxides, carbides, and solid lubricants such as graphite and MoS2 are the common reinforcements used in MMCs.

### *3.1.2 Polymer matrix composites (PMC)*

Polymer matrix composites are composed of thermoplastic or thermosetting plastic as matrix with one or more reinforcements such as glass, steel, carbon and natural fibers.

### *3.1.3 Ceramic matrix composites (CMC)*

Ceramic matrix composites uses ceramic material as matrix and fibers embedded on the matrix made of other ceramic material **Figure 4**.

## 3.2 Advantages of MMC over metals

1. Lighter density and better mechanical properties with an increase in strength per unit weight

2. Better dimensional stability during sintering

3. Better creep resistance with higher-temperature withstanding capabilities

## 3.3 Advantages of MMC over PMC

1. Can service better at high-temperature environments.

2. Electrical and thermal conductivity is higher compared to PMC.

3. It can transfer the load better in transverse direction, by which better properties in transverse direction are achievable and have better joining characteristics.

4. Survives better when subjected to radiation such as nuclear, laser, and UV.

5. Improved fatigue resistance.

## 4. Graphene as a reinforcement

Since 2004, graphene is treated as one of the most wonderful achievements in the field of science and technology Graphene is an attractive alternative to other carbon nanofillers (CNT, CNF, etc.). From the mechanical point of view surface engineering is also meant to be considered an attractive instrument for tribological challenges, several solutions involving chemical, structural and morphological modification by means of adding graphene on surface topography, can be adopted with the aim of improving performances, reducing friction and wear, and/or increasing hardness and toughness [15]. Latest trend has shown that texturing the surfaces of cutting tools can have a beneficial influence on the tribological properties when cutting different types of materials [16]. Moreover, surface textures can act as reservoirs for cutting fluid and increase the fluid's retention on the tool surface leading to enhanced lubrication and cooling. Surface texturing includes modifications of exterior geography, making a uniform smaller scale help with consistently formed severities or depressions. Latest research has considered an advancement of small-scale dimples for automobile portions including bearing and cylinder rings. This examination highlighted the significance of different examples and its impact on decreasing the friction coefficient [17].

Graphene is a novel material in today's scientific world and possesses some excellent properties, such as large specific area, two-dimensional high aspect ratio sheet geometry, and outstanding mechanical, electrical, and thermal properties [18]. The hexagonal crystalline single layer of graphite (the simplest form and one of the most important crystalline allotropes of carbon atoms having a C—C bond distance of 0.142 nm) has received massive attention in the field of sensors, bio-medicals, composite materials and microelectronics. Thus, graphene is the strongest and best conductive material [19]. Graphene platelets or multi-layer graphene has been used as toughening additives to enhance the mechanical properties of various cutting tool materials [20]. Graphene and carbon nanotubes have attracted much attention in recent years due to their extremely high thermal conductivity, strength and exceptional tribological behavior. Graphene possesses similar mechanical properties as CNTs but has superior electrical and thermal properties, and a larger surface area (2620 $m^2/g$) [21] because of its 2-dimensional crystal structure. High strength, high thermal properties, and tribological behavior make graphene a good candidate as a reinforcement material for MMCs. Still, study of WC-Co-based cutting tool material made by graphene powder is quite a few **Table 1**.

| Density | Melting point | Young's modulus | Coefficient of friction | Thermal conductivity |
|---|---|---|---|---|
| 2.0 $g/cm^3$ | 4237 °C | 1034 GPa | 0.03 | More than 3000 W/mK |

**Table 1.**
*The properties of multilayer graphene used as reinforcement.*

## 5. Fabrications methods

The development of fabrication processes for the production of high-performance composites has been reported in many research studies [22]. **Figure 5** illustrates various fabrication techniques that were in use for the last few years. The following methods are most common for fabrication of the MMCs at large-scale industrial level.

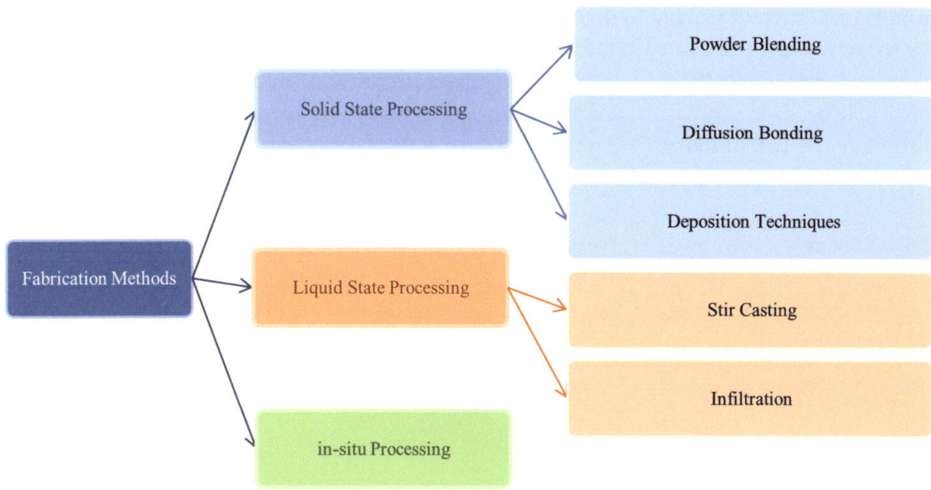


**Figure 5.**
*Different fabrication method of composite cutting tool.*

In the past decade, there has been a research done on the fabrication of MMCs with the different proportions of reinforcement for obtaining the required properties.

Liquid-state processing is defined as the incorporation of reinforcement in the matrix in molten form to prepare metal matrix composite. The various advantages of liquid-state processing are:

- Faster processing rate, especially when dealing with low melting point alloys of aluminum and magnesium. It can also produce near net shape of the final component in a single step.

Liquid-state processing of metal matrix composites involves incorporating or combining a liquid metal matrix with the reinforcement. The most common liquid-phase processing techniques can be subdivided into four major categories:

- Casting or liquid infiltration
- Squeeze casting or pressure infiltration
- Spray co-deposition
- Infiltration

Solid-state processing is preferred because of the following drawbacks in the liquid-state processing:

- Distribution of reinforcement is difficult to control.
- Achievement of uniform microstructure is difficult.

The most common solid-phase processes are based on powder metallurgy techniques, which generally use discontinuous reinforcements such as flakes/particulates/

short fibers, since it is easy to do the mixing and blend with it and much more effective to reduce the porosity [23]. During the process, the ceramic and metal powders are mixed, are statically cold compacted, sintered, and hot pressed to get higher densification or it may be mixed and blended followed by direct hot isostatic pressing to achieve the full density. Further secondary processing such as extrusion, forging etc., might be required to achieve the final shape of the component.

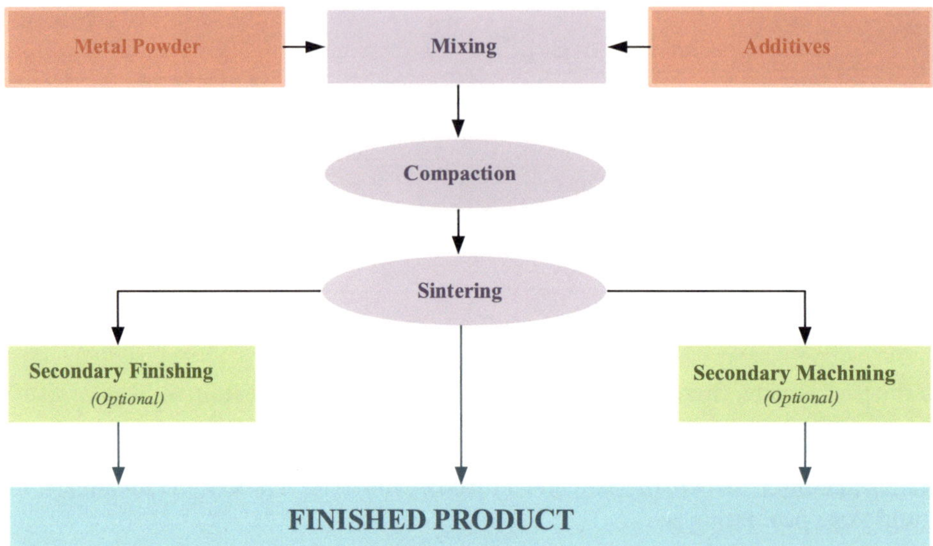


In-situ processing involves chemical reactions that result in the creation of a reinforcing phase within a metal matrix. The reinforcements can be formed from the precipitation in liquid or solid. This method provides thermodynamic compatibility at the matrix reinforcement interface. The reinforcement surfaces are also likely to be free of contamination and, therefore, a stronger matrix-dispersion bond can be achieved. Different researchers reported the benefits of using these techniques such as lower processing time, temperature as well as the capabilities of producing near net shape, high densification rate and less energy required.

## 6. Previous work on fabrication of tungsten carbide-cobalt-based composite

In the literature, various methods have been introduced such as coating, MQL, cryogenics, microwave heating, and high-pressure coolant cutting to improve the performance of various cutting tools.

Arshi [24] have studied to improve surface roughness, tool life and production rate of materials by using a protective covering of titanium nitride (TiN) over steel tools. They have obtained satisfactory results. Huang et al. [25] manufactured WC powder was prepared by the SPS process at 1873K, 8 min, 60 MPa pressure without the addition of any binder phase. Results obtained that binderless WC sintered at 1773 K for 4 min showed almost full densification with a relative density of 99.6% and higher Vickers hardness up to 2600 HV, compare to conventional WC-Co cemented carbides. Yuchi et al. [26] investigated fully dense GNS/$Al_2O_3$ composites fabricated from ball-milled and then expanded in a graphite die by using SPS. The result showed that conductivity increased when composite has 15% GNSs volume which was 170% higher than CNT/$Al_2O_3$ composites. Virendra et al. [27] studied graphene-based materials' impact on electronic devices, chemical sensors, nano-composites and energy storage. Various synthesis processes of single-layer graphene, graphene nano-ribbons, chemically derived graphene, and graphene-based polymer and nano-particle composites are reviewed. Prashantha et al. [28] have shown that graphene has noble mechanical properties, which makes it good alternative reinforcement in metal matrix composite. He has also focused on various dispersion methods, mechanisms of strengthening, composites synthesized using graphene, and their applications.

Meanwhile Bashirvand and Montazeri [29] made metal-based composite supported with carbon nanofiller such as graphene sheets and carbon nanotubes (CNTs) were proposed to join the properties of metals. These nanofillers have prompted novel materials for different applications. The outcomes showed that under the same conditions, graphene sheets performed altogether better compared to CNTs to improve thermo-mechanical. Dongguo et al. [30] manufactured a cutting tool by powder injection molding (PIM) technology using 90WC-10Co alloy. The cutting tool obtained after this technology achieved a density of 95% of theoretical density and dimension accuracy achieved was 98%. T. Wejrzanowski et al. [31] introduced the advantages and limitations of applications of single-layer graphene (SLG) and multi-layer (MLG) graphene for thermal conductivity enhancement (TCE) of copper and showed that the volume fraction of multi-layer graphene, their size, distribution and orientation may significantly affect the thermal conductivity of metal matrix composites. Das et al. [32] studied that graphene attracts particular interest due to its novel properties like high thermal conductivity, high mechanical strength and self-lubricating properties etc. Ali Nasser et al. [7] used another methodology for stabilizing a predesigned Co gradient in the microstructure of nano-WC-Co thinning structure via graphene additions is presented. For this purpose, laminated specimens of green WC-Co functionally graded material, having three layers structured, with and without graphene additions in the intermediate layer were sintered at solid and liquid sintering temperatures of 1290 and 1400°C, respectively, using the hot isostatic pressing technique (HIP).

Grasso et al. [33] demonstrated the densification of high-purity nano-structured tungsten carbide by High-Pressure Spark Plasma Sintering (HPSPS) in the unusually low-temperature range of 1200–1400°C. The high-pressure sintering up to 300 MPa produced dense material at a temperature as low as 1400°C. In comparison with more conventional sintering techniques, such as SPS (80 MPa) or hot isostatic pressing, HPSPS lowered the temperature required for full densification by 400–500°C. Bódis et al. [34] prepared silicon carbide (SiC) ceramics that have superior properties in terms of wear, corrosion, oxidation, thermal shock resistance, and high-temperature mechanical behavior, as well. In this work, SiC-based ceramics mixed with 1 wt% and 3 wt% multilayer graphene (MLG), were fabricated by solid-state spark plasma sintering (SPS) at different temperatures. It was found that MLG improved the mechanical properties of SiC-based composites due to the formation of a special microstructure. In other addition of 3 wt% MLG to SiC matrix increased the Vickers hardness and Young's modulus of composite, even at a sintering temperature of 1700°C. Suna et al. [35] investigated mechanical and tribological properties of functionally graded multilayer graphene (MLG)-reinforced WC-TiC-$Al_2O_3$ ceramics prepared to employ two-step sintering (TSS) are determined in this paper. Results showed that MLG can act as not only an outstanding reinforcement phase but also act as self-lubricant phase. As result demonstrated that 0.1wt% of MLG/WC-TiC-$Al_2O_3$ ceramics exhibit 53.3% enhancement in fracture toughness, 73.8% decrement in friction coefficient, 82.65% improvement in wear resistance in comparison with monolithic ceramics. The study of Gorti et al. [36] revealed that graphene as reinforcement could be applied in cemented carbides. WC powder with 6% cobalt (Co) and graphene (0.2%) in the form of graphene nanoplatelets (GPLs) was set up by high energy rate ball milling and ultra-sonification. After that mixture was sintered by utilizing spark plasma sintering at 1250°C for 10 min. Results found that spark plasma sintering of graphene reinforced WC-Co composite resulted in a significant increase in toughness. It gave higher hardness (400 Hv) and it makes the grain size conveyance smaller. From different studies, it is clear that the reaction of graphene is limited in spark plasma sintering (SPS) as compared to Hot isostatic processing (HIP) and Graphene goes about protective coating against oxidation.

Karthikeyan et al. [37] studied the effect of laser surface textured tungsten carbide (WC-Co) insert and filled with graphite, which helps in reducing chip adhesion during machining of aluminum AA2025 studied and the following conclusions were carried out. A tribological test was carried out to investigate the frictional behavior of untextured, textured and textured inserts filled with graphite powder. The outcome clearly showed that the coefficient of friction between work material and textured graphite filled inserts reduced approximately by 12% and 90%, respectively, when compared with untextured and textured inserts. Singh [38] proposed the near rake face cutting edge of carbide turning insert were polished and used graphene as a potential solid lubricant was applied. It was found that cutting forces and coefficient of friction at tool-chip interface diminished significantly while turning with inserts applied with graphene. It was also clear that the effect of graphene on tool-chip interface significantly decreased to almost negligible when main cutting force increased beyond 60N. Hence more sincere research and development efforts are required to make its use sustainable in machining as a solid lubricant.

Durwesh et al. [39] aimed to move toward good product quality and better productivity. We know that adverse machining conditions result in fast tool wear, a decrease in surface finish, and an increase in cutting forces. Results demonstrate that microwave-irradiated tool inserts perform better during machining of AISI 1040 steel when contrasted with uncoated inserts. The result indicated that 30.2% increase in tool hardness was observed in 30-min microwave-treated insert and tool wear was reduced by 25–35%. Chen et al. [40] presented the effect graphene and carbon nanotubes were blended with WC-Co powder and sintered by spark plasma sintering technique (SPS). The outcomes showed that adding a small amount of graphene or carbon nanotubes helped to increase the bending strength of the cemented carbide by approximately 50% while keeping the hardness of the cemented carbide constant & thermal conductivity of the cemented carbide has also increased by 10% with the addition of 0.12 wt% graphene [41]. Virendra Singh et al. [42] described that higher mechanical properties (elastic modulus and tensile strength) of graphene sheets have attracted the attention of researchers. Vandana et al. [43] investigated and talked about the addition of graphene to $Al_2O_3$ ceramic matrix and its effect on different mechanical properties of resulting alumina-graphene (Al-G) composite tool material. The wt% of graphene varied from 0.15 to 0.65 with an interval of 0.1%. The result showed that composite with 0.45 wt% of graphene yielded the maximum hardness and fracture toughness. Lagos et al. [44] introduced the changes in densification behavior and mechanical properties of $Ti_3SiC_2$ composites containing 0–40 volume % of short carbon fibers densified by Spark Plasma Sintering Technique (SPS). It was feasible to obtain fully densified composites up to 20 volume % of carbon fibers and more than 90% of the theoretical density with the 40 volume % of fibers.

Zhenhua et al. [45] examined ultrafine-grained WC-12Co-0.2VC cemented carbides prepared by using two-step spark plasma sintering (SPS) technique. Thus, the first-step (T1) and the second-step (T2) temperatures in the two-step SPS are 1300°C and 1200°C, respectively. He has talked about the effect of the holding time during the first and second steps on the mechanical properties of the specimen. The results showed that the UYG12V cemented carbide sintered at 1300°C for 3 min and then at 1200°C for 5 min has the best extensive mechanical properties, Vickers hardness, fracture toughness, relative density, and bending strength of 218.06 GPa, 12.25 MPa $m^{1/2}$, 99.49%, and 1960 MPa, respectively. Xuchao et al. [46] chose graphene as reinforcement in $Al_2O_3$-WC-TiC composite ceramic tool materials by hot pressing technique. The optimal flexural strength, Vickers hardness, and indentation fracture toughness were 646.31 ± 20.78 MPa, 24.64 ± 0.42 GPa, 9.42 ± 0.40 MPa $m^{1/2}$, respectively, at 0.5 volume % of graphene content, which was significantly improved compared to ceramic tool material without graphene. Yuchi et al. [47] studied to obtain fully dense GNS/$Al_2O_3$ composites have been fabricated from ball-milled graphite and $Al_2O_3$ by spark plasma sintering (SPS). The GNSs after ball processing are 2.5–20 nm in thickness and homogeneously dispersed in the ceramic matrix. The conductivity achieves 5709 S/m when composite has 15% volume GNS, which was 170% higher contrasted with the best outcome recently announced in CNT/$Al_2O_3$ composites. Yanju et al. [48] created ultrafine cemented carbides were set up by microwave sintering technique using WC-V8C7-$Cr_3C_2$-Co nano-composites as a raw material. The outcomes showed that the ultrafine solidified carbides arranged at 1300°C for 60 min have better mechanical properties. The relative density, Vickers hardness, and fracture toughness of the composite reach the maximum values of 99.79%, 1842 kg/$mm^2$ and 12.6 MPa $m^{1/2}$ respectively.

After studied the above literature it has observed that cutting tool made with different type of reinforcements at different manufacturing condition play a very important role on the performance of composite cutting tool. Different composite like GNS/$Al_2O_3$, CNT/$Al_2O_3$, SiC-MLG, WC-TiC-$Al_2O_3$, WC-CO-GPLs, Al-Gr, and $Al_2O_3$-WC-TiC have prepared by different consolidation techniques and they reported a beneficial influence of all these reinforcement in mechanical, densification, and thermal properties. Further, more research investigation will also be carried in future to enhance the properties, promoting the application and commercialization of improved cutting tool bit for conventional machining.

## 7. Processing challenges

There are different parameters that have their roles in determining the final properties of composites, e.g., starting powder size selection, morphology of starting powders, choosing the appropriate volume fractions, various processing techniques, consolidation techniques, and sintering. All the above-mentioned properties play very important roles in fabricating cutting tool, but temperature, holding time, and pressure have major roles in spark plasma sintering technique to fabricate composite cutting tool. **Table 2** showed the study of various composite cutting tool with different percentage of graphene at different temperature, holding time, and pressure, and their effect on microstructure, densification, physical, mechanical properties, and thermal conductivity have been reviewed.

| Sr. no | Matrix & reinforcement | Graphene | Temperature | Consolidation process | Pressure | Holding time | Result | Year & ref no. |
|---|---|---|---|---|---|---|---|---|
| 1 | WC | — | 1573–1873 K | SPS | 60MPa | 8 min | Vickers hardness 2600 Hv Fracture toughness 9–15 MPa $m^{1/2}$ | 2005 &B. Haung et al. [25] |
| 2 | $Al_2O_3$ | GNS | 1300 °C | SPS | 60 MPa | 3 min | Conductivity achieves 5709S/m | 2010 & Fan et al. [26] |
| 3 | WC | — | 1400°C | HPSPS | 300 MPa | 10 min | (2721 HV10) 7.2 MPa $m^{1/2}$ | 2013 & Grasso et al. [33] |
| 4 | WC-Co | | 50°C | Powder injection molding | 37 MPa, | 0.6 s | Final density of 14.21 $g/cm^3$ | 2015 & Lin et al. [30] |
| 5 | Mullite ($3Al_2O_3.2SiO_2$) | WC (5-20)% | 1700°C | SPS | 30 MPa | 3 min | Maximum Strength 298 MPa and Vickers hardness 1589 HV | 2016 & H. Rajaei et al. [49] |
| 6 | Steel | $TiB_2$ | (1000°C & 1100°C ) | SPS | 35 MPa | 5 min & 30 min | Both compression strength & hardness increased | 2015 & Sulima et al. [35] |
| 7 | WC | GNP (0.5–6 wt.%) | 1850°C | HP | 25 MPa | 60 min | Relative density 98% | 2016 & Kornaus et al. [10] |
| 8 | WC-TiC-$Al_2O_3$ | 0.1 wt% MLG | 1800°C | TSS | — | — | Toughness 14.5 MPa $m^{1/2}$ | 2017 & Sun et al. [50] |
| 9 | Silicon carbide (SiC) | 3 wt% MLG | 1700 °C | SPS | 50 MPa | 5 min | Fracture toughness increase by 20% & hardness also increase | 2017 & Bodies et al. [34] |
| 10 | TiB2/TiC | 0.5 wt.% GNP | 1750°C | SPS | 30 MPa | 5 min | Fracture toughness 8.3 ± 0.43 MPa Hardness 19.3 ± 0.52 GPa | 2018 & Yin et al. [51] |
| 11 | $Al_2O_3$-WC-TiC | 0.25% | 1700°C | HP | 35 MPa | 10 min | Vickers hardness, 646.31 ± 20.78 MPa, 24.64 ± 0.42 GPa fracture toughness, 9.42 ± 0.40 MPa $m^{1/2}$ | 2019 & Wang et al. [26] |
| 12 | WC-Co | MLG | 1400°C | TSS | | 5 min | Relative density 99.8% fracture toughness 15.9 ± 1.2 $m^{1/2}$ | 2018 & Sun et al. [42] |
| 13 | WC-CO | (0.2%)GPL | 1250 °C | SPS | 50 MPa | 10 min | Hardness 2000 fracture Toughness 11.8 | 2018 & Kiran et al. [36] |
| 14 | $Al_2O_3$ | 0.15–0.65 | 1500°C | Microwave furnace | | 30 min | Fracture toughness2.68 MPa $m^{1/2}$ | 2019 & Vandana et al. [43] |

| Sr. no | Matrix & reinforcement | Graphene | Temperature | Consolidation process | Pressure | Holding time | Result | Year & ref no. |
|---|---|---|---|---|---|---|---|---|
| 15 | WC–Co | 0.12 wt% | 1200°C | SPS | 80 MPa | 10 min | Thermal conductivity 9.86% | 2019 & Chen et al. [40] |
| 16 | WC-12Co-0.2VC | — | 1100°C | SPS | 30 MPa | 5 min | Vickers hardness, 18.06 GPa, fracture toughness12.25 MPa $m^{1/2}$ | 2019 & Wang et al. [45] |
| 17 | TiC | 0.45 wt% MLG | 1850°C | TSS | 40 MPa | 60min | Hardness 24.36 GPa, flexural strength 708.9 MPa fracture toughness of 7.28 MPa $mm^{1/2}$ | 2020 & Jialin Sun et al. [52] |

**Table 2.**
*Consolidation of different composite materials by various techniques.*

Graphene, due to its unique combination of electrical, mechanical and thermal properties can greatly improve simultaneously properties of obtained composites. Addition of graphene seems a new idea is to obtain self-lubricating carbide cutting tool materials with the addition of various graphene reinforcements.

## 8. Conclusions

This paper gives a state of art and recent development toward various techniques to extend the lifetime of cutting tool and to improve the mechanical and tribological properties of the material. Various study revealed that graphene attracts particular interests in latest for cutting tool development due to its unique properties like high temperature operations, high mechanical strength, high thermal conductivity, and its self-lubricating properties. Previous studied has also showed that properties and microstructure of the composite depend upon strongly SPS conditions. Mostly, graphite is used as solid lubricant to improve the efficiency of various carbide tool materials; however, the study of multilayer graphene to make carbide tools as self-lubricant was relatively few. Results showed that MLG can act as not only an exceptional reinforcement phase but also act as superior self-lubricant phase. Very few Investigations have done on Graphite as reinforcement. Since not much research was done on machining and tribological analysis of fabricate cutting tool insert so far, there are lot of aspects which require improvement. More work should be done to obtain composites with higher homogeneity, which will result in fabricating samples with even better mechanical properties and higher density. Spark Plasma Sintering process seems good method for sintering powders where no degradation to graphene is required. It is strongly recommended that more in-depth studies will be carried out with regard to the preparation of composite, which leads to better mechanical parameters for obtained composites, comparable with commercial insert.

## Author details

Gaurav Saini[1*], Sarabjeet Singh Sidhu[2] and Simranpreet Singh Gill[2]

1 Department of Mechanical Engineering, Aman Bhalla Institute of Engineering and Technology, Pathankot, India

2 Department of Mechanical Engineering, Sardar Beant Singh State University, Gurdaspur, India

*Address all correspondence to: gauravsaini.saini19@gmail.com

## References

[1] Chandrashekar BH, Gowda AC, Udaykumar PA. Study on Cutting Tool Life. Journal of Engineering and Science (IJES). 2014;**2**:2319

[2] Badaluddin NA, Zamri WFHW, Din MFM, Mohamed IF, Ghani JA. Coatings of cutting tools and their contribution to improve mechanical properties: A brief review. International Journal of Applied Engineering Research. 2018;**13**(14):11653-11664

[3] Chavan AA, Deshmukh PV. Prediction of tool life of different coated cutting toolsduring machining of Inconel 718. International Research Journal of Engineering and Technology. 2017;**12**(4):941-949

[4] Kaam SV, Rath MG. Review of different approaches to improve tool life. International Journal of Innovative Research in Science Engineering and Technology. 2014;**3**:514-521

[5] Narasimha M, Sridhar K, Reji Kumar R. Improving cutting tool life a review. International Journal of Engineering Research and Development. 2013;**7**:67-75

[6] Mahaidin AA, Jaafar TR, Selamat MA, Budin S, Sulaiman ZS. Effect of fabrication process on physical and mechanical properties of tungsten. American Institute of Physics. 2017;**1901**:040006-1-040006-5

[7] Ali N, Kazunari S, Hassan E-H, Moneim AA. Enhancing stability of Co gradient in nano-structured WC–Co functionally graded composites using graphene additives. Journal of the Ceramic Society of Japan. 2016;**12**(124):1191-1198

[8] Rizzo A, Goel S. The critical raw materials in cutting tools for machining applications: A review. Materials. 2020;**6**:1-48

[9] García J, Collado V, Andreas C. Cemented carbide microstructures: A review. International Journal of Refractory Metals and Hard Materials. 2022:40-68

[10] Kornaus K, Gubernat A, Zientara D, Rutkowski P, Stobierski L. Mechanical and thermal properties of tungsten carbide-graphite. Polish Journal of Chemical Technology. 2016;**2**(18):84-88

[11] Grigoriev SN, Fedorov SV, Hamdy K. Materials, properties, manufacturing methods and cutting performance of innovative ceramic cutting tools. A Review Manufacturing Review. 2019;**6**(19):1-27

[12] Kellen D. Traxel and amit bandyopadhyay stellite™ cutting tools using laser-based additive manufacturing. Additive Manufacturing. 2019;**25**:460-468

[13] Nagavally RR. Composite materials—history, types, fabrication techniques, advantages, and applications. International Journal of Mechanical and Production Engineering. 2017;**5**:82-87

[14] Sahu PS, Banchhor R. Fabrication methods used to prepare Al metal matrix composites—A review. International Research Journal of Engineering and Technology. 2016;**3**(10):123-132

[15] Nassef BG, Nassef GA, Mohamed A. Daha graphene and its industrial applications — A review. International Journal of Materials Engineering. 2020;**10**:1-12

[16] Geim AK, Novoselov KS. The rise of graphene. Nature Materials. 2007;**6**(3):183-191

[17] Lin C-R. Effects of surface modification on improvement of diamond coating on tungsten carbide cutting tool. International Journal of Mechanical Engineering and Robotics Research. 2016;**3**:168-170

[18] Singh R. Wear behavior of textured tools under graphene-assisted minimum quantity lubrication system in machining Ti-6al-4v Alloy. Tribology International. 2020;**145**

[19] Kornaus K, Gubernat A, Zientara D, Rutkowski P, Stobierski L. Mechanical and thermal properties of tungsten carbide – graphite. Polish Journal of Chemical Technology. 2016

[20] Nieto A, Bisht A, Lahiri D, Zhang C, Agarwal A. Graphene reinforced metal and ceramic matrix composites: A review. International Materials Reviews. 2016;**62**:1-63

[21] Suna J, Zhaoa J, Chenb M, Zhoua Y, Nia X, Lia Z, et al. Multilayer graphene reinforced functionally graded tungsten carbide. Materials and Design. 2017;**134**:171-180

[22] Kandpal BC, Kumar J, Singh H. Production technologies of metal matrix composite: A review. International Journal of Research in Mechanical Engineering & Technology. 2014;**4**:27-31

[23] Prashantha Kumar H, Ga M, Xavior A. Graphene reinforced metal matrix composite (GRMMC): A review. 2014;**97**:1033-1040

[24] Arshi. Effect of SPS parameters on densification and properties of steel matrix. Advanced Powder Technology. 2015;**4**(26):1152-1161

[25] Huang B, Chen LD, Bai SQ. Bulk ultrafine binderless WC prepared by spark plasma sintering. Scripta Materialia. 2006;**3**(54):441-445

[26] Yuchi Fan A, Lianjun Wang C, Li J, Jiaqi Li A, Shikuan Sun A, Feng Chen A, et al. Preparation and electrical properties of graphene nanosheet/$Al_2O_3$ Composites. Carbon. 2010;**48**:1743-1749

[27] Singh V, Joung D, Zhai L, Soumen Das AB, Khondaker SI, Sudipta Seal A. Progress in materials science. Graphene Based Materials: Past, Present and Future. 2011:1178-1271

[28] Sun J, Zhao J, Gong F. Design, fabrication and characterization of multi-layer graphene. Journal of Alloys and Compounds. 2018;**750**:972-979

[29] Bashirvand S, Montazeri B. New aspects on the metal reinforcement by carbon nanofillers: A molecular dynamics study. Materials and Design. 2016;**91**:306-313

[30] Lin D, Jifu X, Shan Z, Chung ST, Park SJ. Fabrication of WC-Co cutting tool by powder injection. International Journal of Precision Engineering and Manufacturing. 2015;**7**(16):1435-1439

[31] Wejrzanowski TA, Grybczuk MA, Chmielewski MB, Pietrzak KB, Kurzydlowski KJA, Strojny-Nedza AB. Thermal conductivity of metal-graphene composites. Materials and Design. 2016;**99**:163-173

[32] Das DK, Santra S, Sahoo S. Graphene coating on cutting tools can remove the use of coolants. Journal of Nanoscience, Nanoengineering & Applications. 2016;**6**:1-5

[33] Grasso S, Poetschke J, Richter V, Maizza G. Low-temperature spark plasma sintering of pure nano WC powder. Rapid Communications of The American Ceramic Society. 2013:1701-1705

[34] Bódis E, Cora I, Balázsi C, Németh P, Károly Z, Klébert S. Spark plasma sintering of graphene reinforced silicon carbide. Ceramics International; 2017(1-7);**43**(12):9005-9011

[35] Suna J, Jun Zhao B, Huang Z, Yanc K, Fei Chen C, Yongxin Jian A, et al. Preparation and properties of multilayer graphene reinforced binderless Tic nanocomposite cemented carbide through two-step sintering. Materials and Design. 2020;**188**:1-8

[36] Kiran GV, Golla Sairam B, Shankar B, Jyothi K, Jhansi K, Rao RS, et al. Fabrication and characterization of graphene reinforced tungsten carbide-cobalt composite. The Journal of Advances in Mechanical and Materials Engineering. 2018;**1**:12-22

[37] Karthikeyan S, Kannan S, Ravi Kumar K. Impact of laser micro textured cutting tool inserts filled with graphite in reducing chip adhesion during the turning process of aluminium alloy AA2025. International Journal of Mechanical and Production Engineering Research and Development. 2018:1285-1292

[38] Singh S. Preliminary investigation on graphene as solid lubricant in dry turning. International Journal of Advance Research in Science and Engineering. 2018:2331-2337

[39] Jhodkar D, Amarnath H, Chelladurai JR. Experimental investigations to enhance the machining performance. Journal of the Brazilian Society of Mechanical Sciences and Engineering. 2017;**40**:2-16

[40] Chen WX, Li Z, Wu J, Hong K, Ruan X. Effect of graphene and carbon nanotubes on the thermal conductivity of WC–Co cemented carbide. Metals. 2019:2-14

[41] Pydi HPR, Adhithan B, Bakrudeen ASB. Microstructure exploration of the aluminum-tungsten carbide composite with different manufacturing circumstances. International Journal of Soft Computing and Engineering. 2013;**2**(6):257-261

[42] Singh V, Joung D, Zhai L, Soumen Das AB, Khondaker SI, Sudipta Seal A. Progress in materials science. Graphene Based Materials: Past, Present and Future. 2011;**18**:1178-1271

[43] Vishnu Vandana KI, Suman KNS. Hardness and fracture toughness of ceramic composite using experimental and analytical methods. International Journal of Engineering and Advanced Technology. 2019;**17**:5250-5254

[44] Lagosa MA, Pellegrinib C, Agotea I, Azurmendia N, Barcenaa J, Parcoa M, et al. $Ti_3SiC_2$-Cf composites by spark plasma sintering: Processing, microstructure. Journal of the European Ceramic Society. 2019;**39**:2824-2830

[45] Wang Z, Jia J, Wang B, Wang Y. Two-step spark plasma sintering process of ultrafine. Materials. 2019;**12**:1-11

[46] Xuchao Wanga B, Jun Zhaoa B, Enzhao Cuia B, Songa S, Liua H, Songa W. Microstructure, mechanical properties and toughening mechanisms graphene reinforced $Al_2O_3$- WC-TiC composite ceramic tool material. Ceramics International. 2019:10321-10329

[47] Yuchi Fan A, Lianjun Wang C, Li J, Jiaqi Li A, Shikuan Sun A, Feng Chen A, et al. Preparation and electrical properties of graphene nanosheet/$Al_2O_3$ composites. Carbon. 2010;**48**:1743-1749

[48] Qian Y, Zhao Z. Microstructure and properties of ultrafine cemented carbides prepared. Crystals. 2020;**10**(6):507

[49] Iwona Sulima A, Piotr Putyra B, Paweł Hyjek A, Tomasz TC. Effect of SPS parameters on densification and properties of steel matrix. Advanced Powder Technology. 2015;**26**:1-10

[50] Jialin Sun AB, Jun Zhao AB, Feng Gong AB, Zuoli Li AB, Xiuying Ni AB. Design, fabrication and characterization of multi-layer graphene. Journal of Alloys and Compounds. 2018:972-979

[51] Suna J, Zhaoa J, Chenb M, Zhoua Y, Nia X, Lia Z, et al. Multilayer graphene reinforced functionally graded tungsten carbide. Materials & Design. 2017;**134**:1-28

[52] Yin Z, Yuan J, Xu W, Chen M, Yan S, Wang Z. Effect of Ni and graphene on microstructure and toughness of titanium boride ceramic tool material prepared by spark plasma sintering. Ceramics International. 2018;**44**:1-11

Chapter 12

# Graphene Oxide Based on Biomass Waste: Synthesis and Applications

*Ramli Ramli and Rahmat Hidayat*


## Abstract

Graphene oxide is a two-dimensional material formed from oxidized graphite, with oxygen (O) functional groups decorating the *sp*2 plane of carbon (C). Graphene oxide can be obtained by exfoliating the graphite oxide (oxidized three-dimensional carbon-based material) into the layered sheets by sonication or mechanical stirring. Graphene oxide contains various reactive oxygen functional groups, which make it to be a good candidate as a foundation in many applications, such as polymer composites, materials for energy conversion, environmental applications, sensors, FET transistors, and photonic applications, due to its excellent electrical, mechanical, and thermal properties. The widely used technique to synthesize graphene oxide is the modified Hummer's method because of its simple process, low cost, and high yield. In this chapter, we report the progress of graphene oxide synthesis using graphite from activated carbon gathered from biomass waste as the source instead of commercial graphite. The chapter covers the synthesis of biomass waste-based graphene oxide and future perspective applications of graphene oxide. Scientific reports about biomass waste-based graphene oxide synthesis and recent applications of graphene oxide will be discussed. The main motivation for writing this chapter is to bring to the horizon the utilization of biomass waste as an alternative carbon source for the green, low-cost, and sustainable production of graphene oxide.


**Keywords:** biomass waste, activated carbon, graphene oxide, Hummer's method

## 1. Introduction

Graphene oxide (GO) is a sheet of graphite oxide obtained by exfoliating graphite oxide into a layered sheet that contains only one or a few layers of carbon atoms through sonication or mechanical stirring [1]. GO can be reduced partially to graphene-like sheets by removing oxygen-containing groups through the restoration of conjugated structure referred to as reduced GO (rGO). These rGO sheets are usually considered to be one type of chemically derived graphene that has similar properties to pure graphene. Graphene and GO are very different, where graphene consists only of sp2 hybridized carbon atoms, while GO has a carbon structure that is decorated by various oxygen functional groups.

The study of GO was reported in 1859 when the English chemist Brodie named it graphite acid or graphite oxide [2], which was prepared by chemical treatment of graphite with potassium chlorate ($KClO_3$) and nitric acid ($HNO_3$). After the graphite oxide has been prepared, GO can be obtained by exfoliating the graphite oxide into


IntechOpen

monolayer sheets through various thermal and mechanical methods [3]. At present, the single-atom carbon layer of graphite oxide is considered graphene oxide (GO).

GO can be produced using inexpensive graphite by applying cost-effective chemical methods with simple processes and high yields. Furthermore, GO is highly hydrophilic and can form stable aqueous colloids to facilitate the assembly of macroscopic structures by a simple and low-cost solution process [4]. The conventional way to convert graphite oxide into GO is carried out by mechanically exfoliating the graphite oxide, by sonication of graphite oxide in water or a polar organic medium into completely exfoliated GO flakes [5]. In addition, through mechanical stirring of graphite oxide in water, graphite oxide can also be well exfoliated into GO [6]. The sonication and mechanical stirring methods can be combined together to exfoliate the graphite oxide producing a better efficiency than the individual methods separately.

It has been reported that there are four methods to synthesize GO including the Brodie method [2], Staudenmaier [7], Hummer and their modifications [8, 9], and Tour method [10]. Nowadays, the synthesis of GO by the modified Hummer's method has become the most common technique for its production. In the modified Hummer method, graphite is served as the main precursor for GO synthesis.

Graphite is classified into natural graphite and synthetic graphite which can be produced by heating the hydrocarbon precursors at very high temperatures. Meanwhile, the combustion of biomass waste produces charcoal which consists of a mixture of hydrocarbons that could be applied as graphite precursors using graphitization process in a lower temperature.

GO contains various reactive oxygen functional groups, which makes it a good candidate for use in many applications, such as polymer composites, materials for energy conversion and environmental applications [11], sensors, FET transistors, as well as biomedical applications, due to its excellent electrical, mechanical and thermal properties [12]. In this chapter, the synthesis of biomass waste-based graphene oxide and the application of GO in electronics, optics, optoelectronics, as well as energy conversion and storage will be discussed.

## 2. Synthesis of graphene oxide from biomass waste

Natural graphite and synthetic graphite have several constraints as the precursors in the production of GO, where the natural sources of graphite are limited in some countries and the production process of synthetic graphite requires extremely high temperatures (≥2500°C) and demands utmostly high cost [13]. Meanwhile, biomass waste has been widely recommended as a potential precursor material for carbon-based synthesis due to its environmental-friendly characteristic, lower temperature process, abundant availability, geographically wide spreading, and lower cost requirements compared to conventional graphite. Several recent studies have suggested biomass as a very appropriate alternative starting material for preparing valuable carbonaceous materials [14–16].

The following section describes the synthesis of GO conducted by the authors by modifying Hummer's method with activated carbon precursors from oil palm empty fruit bunches (OPEFB), rice husks, and coconut shells.

### 2.1 The synthesis of GO from oil palm empty fruit bunches (OPEFB)

Graphene oxide synthesis from OPEFB was started out by drying the collected empty fruit bunches for 2 days under the sunlight to reduce the water content.

Subsequently, the raw material is then chopped into small pieces and put into various 50, 60, 100, and 125 ml evaporating dishes and let dry in the oven for 60 min at 100°C to completely remove water content existing in empty fruit bunches. The raw material the heated in the furnace for 30 min at various temperature of 250, 300, 350, and 400°C, to convert the empty bunches into charcoal. Furthermore, the charcoal is crushed using a mortar and pestle to produce charcoal powder and sieved with a 270 mesh filter.

The carbon activation was carried out by adding the charcoal powder into 50 ml NaOH solution and left the mixture for 24 h. The solution was dried using oven at 105°C for 3 h to obtain the activated carbon sample. **Figure 1** shows XRD pattern of the sample where the diffraction peaks (002) and (100) presents at $2\theta = 29$ degree and $2\theta = 46$ degree indicate the sample could be considered as graphite. The obtained sample was used as a graphene oxide precursor using the modified Hummer's method.

The synthesis of graphene oxide was carried out by mixing 1.5 g of activated charcoal from OPEFB, 0.75 g of $NaNO_3$, and 34.5 ml of $H_2SO_4$ in an Erlenmeyer and stirring the mixture using a magnetic stirrer with 250 rpm at a temperature of 0–5°C for 20 min. The Erlenmeyer was subsequently put into an ice bath to reduce the temperature and keep stirring for 2 h. KMn04 powder was slowly added to the mixture to avoid rapid increase in temperature and explosion. Since 4.5 g of KMn04 was successfully added, Erlenmeyer was removed from ice bath and stirring temperature was set into 35°C for 30 min. The process was carried out until the mixture shows a milk chocolate color. Furthermore, 69 ml of distilled water was slowly poured into the mixture using a dropper and kept stirring for 20 min until the solution color turned dark brown with the appearance of bubbles. The oxidation process was terminated by adding 100 ml of distilled water followed by 1.5 ml of $H_2O_2$ which was indicated by yellow color of the solution. Finally, the solution was diluted by adding 50 ml of distilled water and the graphite oxide sample was gathered. Then, the sample GO was sonicated to peel the graphite oxide into layered graphene oxide. The sample was neutralized by distilled water and centrifuged to separate the precipitate and solvent. The separated precipitate was graphene oxide sample that subsequently dried in the oven to completely remove the water content. The GO sample was confirmed by the characterization instrument. **Figure 2** shows

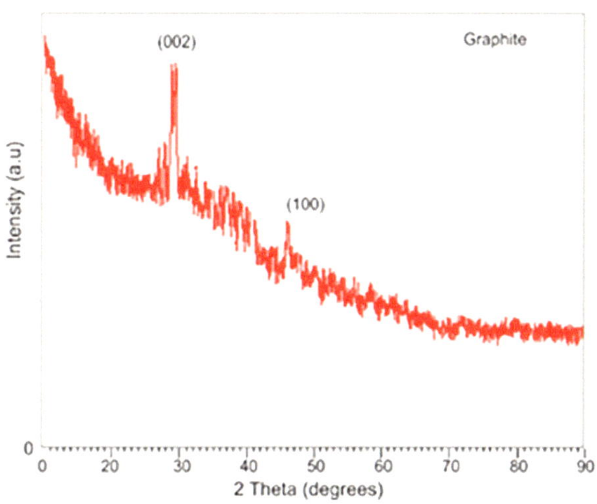


**Figure 1.**
*XRD pattern of activated carbon from (OPEFB).*

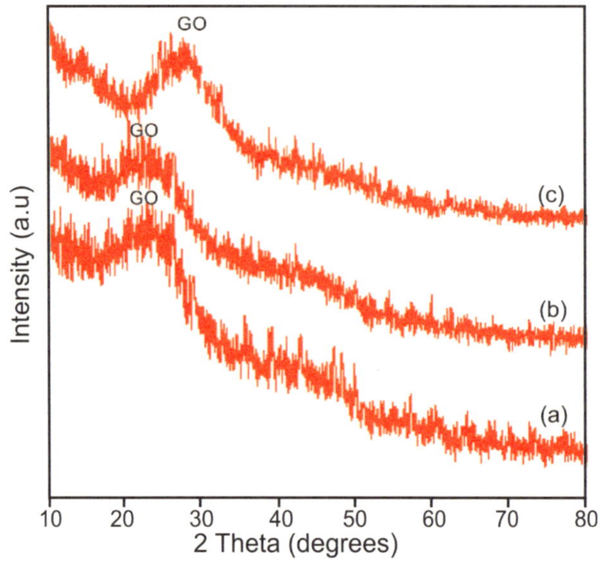


**Figure 2.**
*XRD diffraction pattern of graphene oxide from oil palm empty fruit bunches for carbonization temperature (a) 300 °C, (b) 350 °C, and (c) 400 °C.*

XRD pattern of OPEFB-based GO. Diffraction peak of the sample appears at angle 2θ = 26–29° that indicates the structure of graphene oxide.

## 2.2 The synthesis of GO from rice husk

Rice husk-based GO synthesis began by cleaning the rice husk waste and followed by drying process under sunlight for 3 days. Further drying process was carried out using oven at 105°C for 2.5 h to remove the water and moisture properly. Carbonization process was performed using the furnace for 15 min to produce biocarbon. Various carbonization temperature (250, 300, and 350°C) were used to analyzed the resulted product. Resulted biocarbon was filtered using 140-mesh and 170-mesh sieves to produce uniform particles with a size of 88–106 μm and prepared as the precursor of synthesis process. Modified Hummers method was applied to fabricate rice husk-based graphene oxide. The process was started by mixing 1 g of as prepared rice husk-based biocarbon, 23 ml sulfuric acid ($H_2SO_4$), and various masses of sodium nitrate ($NaNO_3$) in the ice bath and stirred at a speed of 600 rpm for 2.5 h. Various mass of $NaNO_3$ (0, 0.5, 1, and 2 g) were used. Potassium Permanganate ($KMnO_4$) with mass of 3 g was slowly add into the mixture by keeping the temperature under 20°C to prevent explosion and stirred for 30 min. After $KMnO_4$ was completely added, the reaction temperature was raised to 35°C by removing the ice bath and setting the hotplate temperature and stirring process was carried out for 30 min. The mixture was then diluted by slowly adding 46 ml of distilled water and keeping the temperature at 95–99°C and stirring for 30 min. After the oxidation had taken place, diluted Hydrogen peroxide ($H_2O_2$) 3% was added to stop the oxidation process and remove the manganese and permanganate residuals. The oxidation process produced rice husk-based graphite oxide and subsequently, the product was exfoliated to produce rice husk-graphene oxide The solution was set in the Ultrasonic device to perform sonication process. Washing process was performed to neutralize the solution using distilled water and centrifuge for several

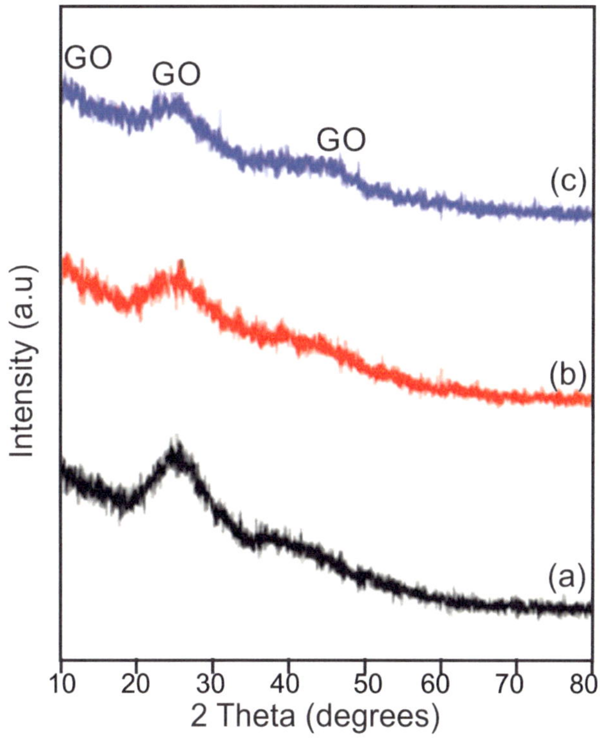


**Figure 3.**
*XRD diffraction pattern of graphene oxide from rice husk waste for carbonization temperature (a) 250 °C, (b) 300 °C, and (c) 350 °C.*

cycles. When the neutral phase was gathered, the precipitate and liquid were separated. As gathered precipitate was dried in the oven at the temperature of 100°C for 15 min and the rice-husk-based graphene oxide sample was obtained.

The XRD pattern of rice husk-based GO is presented in **Figure 3**. All the variations measured have a fairly identical pattern where the diffraction peak appears at the angle of $2\theta = 10°$ with an interplanar distance of 8.8 Å, at the reflection plane (001), and at the angle of $2\theta = 44°$ with an interplanar distance of 2.1 Å, at the reflection plane (100). These characteristics confirmed the formation of graphene oxide. The diffraction peak at about $2\theta = 22°$ indicates the graphene oxide is not completely bound to oxygen atoms. The diffraction peak at $2\theta = 10°$ indicates the distance between the GO layers, while the diffraction peak around $2\theta = 44°$ indicates the short arrangement of the stack layers GO.

The peak between 26° and 44° indicates the presence of an amorphous solid structure that was formed from natural materials. The change in the peak position was influenced by contained oxygen functional groups, which had oxidized graphite to form GO. This result is in accordance with previous researcher's report that the XRD peak of GO nanoparticles from agricultural waste carbonization appears at the angle of $2\theta = 26.6°$ and $2\theta = 44°$.

### 2.3 The synthesis of GO from coconut shells

The synthesis process as began by cleaning the coconut shells and drying under the sunlight for 3 days. Dried coconut shell then was cut into small pieces and heated at 100°C for 60 min to completely remove water and moisture content. Carbonization process of coconut shell was performed using a furnace for 2 h with temperature variations of 250, 300, 350, 400, and 450°C. As gathered coconut shell charcoal was ground and sieved with 125 mesh filter to produce charcoal powder. Subsequently, the charcoal was activated using NaOH solution.

Modified Hummer's method was applied to synthesized coconut shell-based GO. Activated carbon powder from coconut shell with mass of 1.5 g was mixed with 0.75 g Sodium nitrate ($NaNO_3$) and 34.5 ml of Sulfuric acid ($H_2SO_4$ 98%) an Erlenmeyer and stirred for 20 min at a temperature of 0–5°C at a constant speed of 250 rpm. The Erlenmeyer was then put in an ice bath and 4.5 g $KMnO_4$ powder was slowly added by considering the temperature of mixture was below 20°C. The Erlenmeyer was removed from the ice bath and the temperature was increased into 35°C and stirred for 30 min to let the oxidation process take place. Distilled water was added to dilute the mixture by volume of 69 ml and stirring process was continued for 20 min and kept the temperature below 50°C. The mixture showed dark brown color with bubbles. In order to terminate the oxidation process 100 ml of deionized water was added and followed by 1.5 ml of 30% H2O2. The appearance of the mixture turned into yellowish color. The mixture was sonicated for 2 h to exfoliate the graphite oxide into GO followed by washing process using distilled water. The solution was precipitated for 1 day until a liquid and solid phases were formed. Separation of the solid and liquid was carried out using a centrifuge at 4000 rpm for 15 min and followed by GO neutralization. After neutral pH was obtained, GO was dried in the oven at a temperature of 60°C for 12 h.

The coconut shell-based GO sample was measured by XRD characterization that presented in **Figure 4**. Diffraction peak of the sample appears at angle 2θ = 26 and 2θ = 29 degree that signifies the structure of graphene oxide.

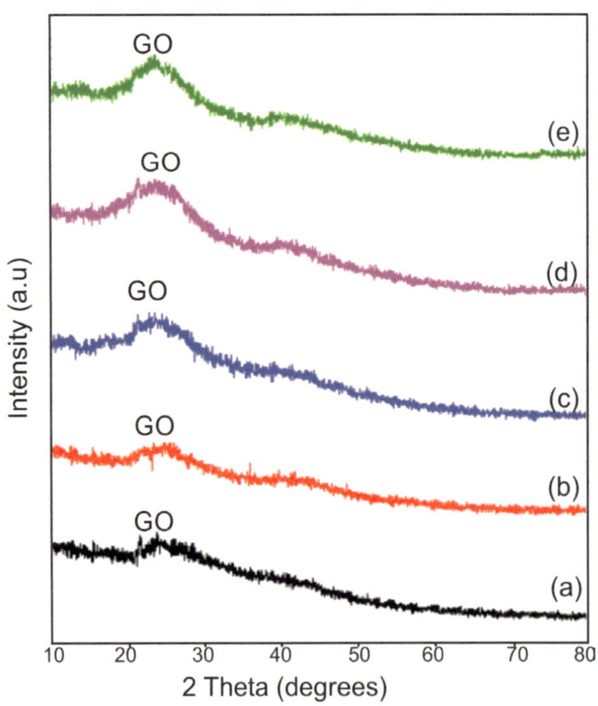


**Figure 4.**
*XRD diffraction pattern of graphene oxide from coconut shell for carbonization temperature (a) 250 °C, (b) 300 °C, (c) 350 °C, (d) 400 °C, and (e) 450 °C.*

## 3. Recent application of graphene oxide

Some recent applications of graphene oxide have been found in various devices and prototypes. In the following section, several recent applications of graphene oxide will be described.

### 3.1 Lithium-ion batteries (LIBs) electrodes

The most widely used power storage in daily gadgets was dominated by a lithium-ion battery which uses graphite material as anode and lithium-based metal oxide as the cathode. GO plays the role in increasing the performance of LIBs by modification of both cathode and anode. GO was added in the fabrication of cathode to modify the property of prior material. The application of GO as additional material in LIBs cathode was found in the manufacture of $LiMnPO_4$ nano-sized cathodes which has been successfully carried out using the solvothermal method. The role of GO in this synthesis is to cut down the $LiMnPO_4$ particle size providing a large surface area, and a short diffusion pathway leading to better performance of a lithium battery. The $Mn^{2+}$ ion coordinates with the oxygen functional group on the GO surface to form a small-sized nucleation site of $LiMnPO_4$. The particle size of $LiMnPO_4$ is inversely proportional to the ratio of GO concentration and precursor solution [17]. The presence of $LiMnPO_4$ on the graphene oxide sheet which is further reduced to reduced GO (rGO) allows a fast electron transfer process between the active material and the collector and accommodates the expansion and contraction of the LiMnPO4 volume during the charging and discharging process [18, 19]. In addition, GO is also used in the synthesis of $LiMn_{0.6}Fe_{0.4}PO_4$ and $V_3O_7.H_2O$ nanorods as the cathode of LIBs. The main role of GO in the synthesis of LIB cathodes is to reduce the particle size so that it reaches nano-size and then GO is reduced using the hydro−/solvothermal method. Apart from being a size reducer, GO is also used to coat cathode materials so as to improve battery performance as found in Vanadium Oxide, $LiFePO_4$, $LiNi_{0.5}Co_{0.2}Mn_{0.3}O_3$ coatings [20, 21].

GO has also been applied intensively in the manufacture of Libs battery anodes. Although GO plays dominantly as an assistant material in the synthesis process, the virtue of GO properties in the anode manufacturing process cannot be neglected. For example, GO was used in the fabrication of GO-assisted $SnO_2$-based anodes in appearance of nanoparticles, quantum dots, spheres and nanorods to improve performance [22–25].

### 3.2 Lithium sulfur batteries electrodes

The nature of GO which has a high surface area and is rich in oxygen functional groups has been utilized in the manufacture of lithium sulfur battery cathodes. The use of sulfur in battery construction is in consideration of cheap, abundant and non-toxic properties. Unfortunately, Sulfur has some lacks in their property such as low electrical conductivity, and the presence of side products during the charge-discharge process creating obstacles in the use of sulfur in batteries. The aim of GO application in the manufacture of Li-S battery cathodes is to accommodate and inhibit S decomposing Li-Polysulfides and to increase the electrochemical stability of Li-S batteries. GO/S nanocomposites can be synthesized by electrostatic self-assembly method. The negatively charged GO sheet is coated on the positively charged S surface [26–28]. The GO/S composite cathode has a capacity of 950–1400 mAh/g and its performance is stable [29].

Various nano-construction with addition of GO provides better results and impact than without GO. The GO-coated porous carbon-sulfur composite-based cathode provides a fading capacity rate of up to 0.12% per cycle for up to 400 cycles [30]. The cathode made of GO/S composite wrapped in amylopectin was also analyzed for its performance against the stability of Li-S battery which confirmed good stability [31]. In another study, GO wrapped in hollow sulfur balls with the aid of polyvinylpyrrolidone resulted in 73% retention at 1C after 150 cycles [32]. The GO-wrapped bowl-like sulfur composites were reported to provide free space to accommodate the volume expansion during press cycling [33]. Other GO-based cathodes such as $MnO_2$-GO [34], cetyltrimethylammonium bromide modified S/GO nanocomposite cathode [35], GO-enclosed walnut-like carbon-based and porous cathode [36], GO and micro-spherical Sulfur composite cathode [37], have been used to increase the performance of Li-S battery with upgrading the capability rate and cycle performance. GO plays the role in the formation of attractive structure of the cathode, increment of the active material loading, and guarantee structural capability during cycling and capture of polysulfides species. In addition, the use of GO with $Li_2S$ is used for long-life and high rate Li-S batteries which provide a decomposition rate of 0.046% per cycle and a coloumbic efficiency of 99.7% for 1500 cycles at 2 C [38].

### 3.3 Supercapacitor electrodes

Supercapacitor is a high-performance electrochemical energy storage that shows excellent properties such as high-power capability, short charge-discharge period, and long cycle life [39]. The charge stored in the SC is principally based on the accumulation of electrostatic charge at the electrode-electrolyte interface which occurs in the electrical double layer capacitance (EDLC), or Faradaic process on the surface of the pseudo-capacitor electrode [40]. GO with outstanding properties has been used to improve SC performance as both electrode and electrolyte. But in the discussion of this book will be limited to the electrodes.

GO can provide pseudo-capacitance and increase the wettability of the electrode in the electrolyte. GO-based electrodes have a larger capacitance than graphene-based electrodes. This experiment was carried out by making GO and graphene electrodes with a mixture of 87 wt% GO samples and graphene, 10 wt% acetylene black and 3 wt% binder which was pressed into pellets. The GO electrode obtained a maximum specific capacitance of 146 F/g and an energy density of 20.39 Wh/kg [41]. Another experiment was reported that $LiCoO_2$-based SC as the positive electrode and GO as the negative electrode and asymmetric SC using porous GO on conductive stainless steel presents the energy density of 19.2 Wh/kg on voltage region up to 1.5 V, and retain specific capacitance of 85% after 1500 cycle [42].

GO can also be used as a template for MnO2 nanosheets to increase the surface area [38] and combined with MnO2 as electrodes to increase capacitance [43–46]. The oxygen group that functions as an anchor site allows the formation of MnO2 nanoparticles that adhere to the surface and ends of the GO sheets. This not only creates a large surface area but also prevents the agglomeration of GO and MnO2. The same thing also happened to GO/$Mn_3O_4$ thin film composite synthesized by the sheet-by-sheet technique giving a specific capacitance of 344 F/g at 5 mV/s [47]. Furthermore, by using the electrophoretic deposition method, or hydrothermal process or co-precipitation, nickel oxide, nickel sulfide and copper oxide can be affixed to the surface of GO nanosheets for high performance SC, namely specific capacitance and current density of 569 F/g at 5 A/g [48], 800 F/g at 1 A/g [49], 245 F/g at 0.1 A/g [50].

Even though the initial electroactive polymers can provide large pseudo-capacitance, but experience swelling properties during the redox process which reduces the stability of the cycle [51]. To solve this problem, GO is widely used to functionalize conductive polymers to make high-performance SCs. The synthesis of conductive GO-polymers can usually be carried out by utilizing electrostatic interactions between negative GO and positive micelles [52]. When the surfactant micelles are electrostatically adsorbed on the GO surface, the addition of polymers and monomers allows the monomers to dissolve in the hydrophobic core of the micelles surfactant and further polymerization processes can take place in the micelles core. For example, GO/Polyaniline (PANI) was synthesized by in situ polymerization [53–56]. Compositing GO and PANI is possible based on three modes namely (1) $\pi-\pi$ stack (2) electrostatic interactions (3) hydrogen bonding [57]. The GO/PANI hybrid has a high surface area and denser pore volume compared to pure PANI. Which is why this can result in a wider contact area between the PANI nanoparticles and with the electrolyte. Which facilitates the rapid transport of electrolyte ions and electrons to the active site of the composite electrode [58, 59]. The synergistic nature of GO/PANI also enhances cycle stability where the GO width maintains the mechanical deformation during the PANI faradaic process which prevents electrode damage.

### 3.4 Transparent electrode

Transparent electrodes are needed in a variety of light-based technologies by passing sunlight into the active layer and at the same time served as electrode for current flow. Transparent electrodes widely used today are made of indium material whose supplies have been running low on the earth. In addition, the use of indium-based materials has is not friendly to the environment and the cost for production is high. The use of carbon-based substitute materials gained most attraction of the scientists to develop this technology [60]. Graphene with all of its virtues is on of carbon based the most attractive candidate for development because it has high electrical conductivity and transparency, both of which are requirements for transparent electrodes. Graphene oxide is produced in a series of processes for making graphene-based transparent electrodes which are then reduced to form reduced graphene oxide (rGO).

Fabrication of rGO thin film on a quartz substrate has been carried out by depositing GO solution using spin-coating method. The process was then followed by a reduction to transform GO into rGO. The deposited rGO thin film showed surface resistance of $10^2$–$10^3$ Ω /sq. and transparency reaching 80% at a wavelength of 550 nm [61]. GO film deposition can also be done in a more convenient way. In another study, facile deposition was carried out with colloidal GO which had been mixed with reductant at low temperature and slowly heated causing the GO is reduced and rGO was self-assembled into glass substrate. Additional treatment was performed by irradiate the sample using commercial microwave oven. The sample has electrical properties of 21.97 kΩ/sq. and optical properties of 30.37% transparency at 550 nm [62].

The transferability of GO to various substrates allows it to be deposited onto flexible substrates. Reduced graphene-based transparent and flexible electrode has applied as moisture sensors attached to human skin. In this case rGO was mixed with polyurethane which gave response and relaxation times of 3.5 and 7 seconds, respectively. This capability does not change when the sensor is stretched to 60% and after 10,000 stretching [63].

### 3.5 Field effect devices

Amazing electrical conductivity induces the potential of GO to be applied as field effect transistor. Good field emission property and bipolar characteristic of its charge carrier allows them to be tuned by using gate electrical field [64, 65]. GO with its oxygen group decoration shows improved field emission of the graphene because the charge concentration and type is very sensitive to dopant. Low threshold field emission of <0.1 V/μm from atomically thin layer of reduced graphene oxide was reported. The edges is decorated by a stable and unique oxygen group of C-O-C bonding from that multiple electron beams are emitted [66].

The field effect emission enables the application of graphene in FET manufacture with more convenient way and low-cost production. Reduced graphene oxide was used to fabricate FET by combination with gold source and dielectrophoresis electrode as drain. When the a backgate voltage was applied, 60% of the devices showed p-type FET behavior, while the remaining 40% showed ambipolar behavior. The experiment was followed by anneal the sample at 200C and giving the result all of ambipolar remains do not change the behavior, meanwhile, the 60% of p-type were change into ambipolar. The maximum charge carrier mobilities in the device are 4.0 $cm^2/Vs$ for hole and 1.5 $cm^2/Vs$ for electron [67].

Further study reports fabrication of Organic FET using GO that grown on a 100 nm $SiO_2$ substrate using doped n-type Si as the gate. GO nano layer was distributed on the substrate by spin coating method at 3000 rpm for 40 s. The following steps diluting PMMA on toluene with concentration of 10 mg/ml and deposited to the substrate by spin coating method. The process followed by anneal the samples under nitrogen atmosphere using a hotplate with temperature of 120 C for 15 min. Deposited GO layer served as charge-tunneling layer. Source and drain electrode are next developed by thermal evaporation under $10^{-7}$ Torr. The performance of Organic FET was analyzed through transfer curves of the sample that shows large gate bias dependent hysteresis with voltage 20 V. After writing an erasing the stored data are well kept with on/off ratio in the order of $10^2$ for $10^4$ s [68].

### 3.6 Electrical sensor

Variation of the oxidation degree or adsorbing molecules can affect the electrical characteristic of GO that allowing it applied as electrical sensor [69]. Reduced GO was found in the fabrication of high-performance molecular sensor by deposit it into substrate to form ultrathin continuous network. The GO film was tunably reduced by varying the exposure time of reductant vapor and reduced GO can served as sensor because of change of conductance after molecular adsorption [70]. GO-based sensor shows lower frequency noise compared to SWNT-based sensor [71].

Other report shows that conductance of reduced GO sensor is depend on the concentration of dopamine molecule. It was revealed that at the gate voltage (Vg) of −0.6 V, the conductance of reduced GO films deposited on flexible polyethylene terephthalate (PET) substrate and concentration of dopamine is directly proportional. In addition, the rGO-based device are able to label-freely detect the hormonal catecholamine molecules and their dynamic secretion from living cells [72].

### 3.7 Flexible electronics

Printed circuit boards face the drawbacks in their use because they cannot be applied to flexible and deformed objects. GO with oxygen functional groups decorating their layers allow it to be deposited onto the various substrates, including flexible substrates. The development of flexible electronics supports the rapid growth of thin, lightweight, and flexible devices. Reduced GO thin films with various thicknesses in rage of one to several layers experiences mechanical flexibility and good electrical conductivity with sheet resistance 43 kΩ/sq. and charge mobility up to 1 and 0.2 $cm^2/Vs$ for electron and hole respectively [73].

Another study reports that GO was self-assembled into platelets and exhibit mechanically flexible, macroporous three-dimensional (3D) properties of carbon films with controllable porous size. Use of nitrogen doping (N-doping) of the 3D assemblies of RGO increase electrical properties and high chemical reactivity. The film firstly transferred onto the SiO2 substrate and further onto flexible PET substrate. The rGO thin film shows sheet resistance of ~128.2 Ω. Further Nitrogen doping decreased the sheet resistances to 13.4 Ω, corresponding to the electrical conductivity of 649 S/cm with the film thickness was 1.15 μm [74].

### 3.8 Photovoltaic devices

Since GO can be used as an fabrication of transparent electrodes, the further application was found in more complex devices such as photovoltaic [75], LED [76], and electrochromic devices [77]. Reduced GO was applied in a dye-sensitized solar cell (DSSC) as the transparent electrode and found that sun light can pass through the electrode and the electrons are injected. A device using fluorine tin oxide (FTO) electrode was also fabricated and evaluated with the same procedure for comparison. The result shows that RGO-based device has lower short-circuit current, which can be attributed to the higher sheet resistance and lower transmittance of the RGO [78].

Application of GO have also been reported for organic photovoltaic (OPV) devices where reduced GO served as electrodes. The reduced GO electrodes were deposited on quartz substrates using spin coating of aqueous dispersion of functionalized graphene, followed by a reduction process. Because of the higher sheet resistance, the short-circuit current and fill factor of these rGO-based device are lower than those of control device on ITO [79]. Flexible transparent conductive electrode-based reduced GO for organic photovoltaic (OPV) was fabricated by transferring chemically reduced GO onto PET substrates. When the transmittance of reduced GO is greater than 65%, the device performance mainly relies on the charge transport efficiency through rGO electrodes, but not sensitive to the transmittance. After the tensile strain (~2.9%) was applied on the fabricated OPV device, it can sustain a thousand cycles of bending Furthermore, the current density of devices can be enhanced by increasing the rGO thickness to lower the sheet resistance, which further improves the overall power conversion efficiency ($\eta$), even if the transmittance of RGO film decreases.

### 3.9 Electrochromic devices

Electrochromic device shows a different appearance in color when some voltage or current was applied. The most widely use of electrochromic device is found on smart glasses, mirrors and windows. Reduced GO/polyaniline (PANI) composite multilayer films were prepared as electrode materials for electrochromic devices to replace conventional indium tin oxide (ITO) The fabrication process carried by deposition of negatively charged graphene oxide (GO) and positively charged PANI upon electrostatic interaction, followed by the reduction of their GO components with hydroiodic acid. The thickness of the multilayer film directly proportional to the number of its bilayers that the thickness of bilayer was about 3 nm. Cyclic voltammetry studies indicated that these thin composite films were electroactive, and their redox reactions were related to the insertion-extraction of counter ions in PANI layers [80].

## 3.10 Photocatalytic water splitting

Photocatalytic water splitting is an artificial photosynthesis process with photocatalysis material used for the dissociation of water molecules ($H_2O$) into hydrogen ($H_2$) and oxygen ($O_2$), using light. The separation of hydrogen and oxygen molecules was utilized for alternatives hydrogen energy, sterilization, anti-fogging, self-cleaning and air purification.

The electronic property of GO depends on the its constituent element that making GO owing p-type behavior because of oxygen's high electronegativity compared to carbon. Likewise, n-type behavior appears when graphene covalently bonds to electron donating nitrogen-containing functional groups [81].

GO was collaborated with other material to enhance photocatalytic performance. It is reported that high solar photocatalytic $H_2$-production activity with reduced GO (RGO)-$Zn_xCd_{1-x}S$ nanocomposite that synthesized by coprecipitation-hydrothermal reduction methode. The optimized rGO-$Zn_{0.8}Cd_{0.2}S$ photocatalyst has a high $H_2$-production rate of 1824 μmol $h^{-1}$ $g^{-1}$ at the RGO concentration of 0.25 wt% and the apparent quantum efficiency of 23.4% at 420 nm. The addition of GO produces improved photocatalytic hydrogen production by 450% compared with that of the pristine Zn0.8Cd0.2S, and better than that of the optimized Pt-Zn0.8Cd0.2S under the same reaction conditions. Reduced GO-Zn0.8Cd0.2S nanocomposite also represents ability to be served as photocatalyts replacing noble metal cocatalysts [82].

## 3.11 Fuel cell

A fuel cell is an electrochemical conversion device that generates electricity from a redox reaction when supplied with fuel such as hydrogen, natural gas methanol or oxidants such as oxygen, air and hydrogen peroxide. GO is applied in the manufacture of fuel cells to increase the electrocatalytic activity. In addition, GO is also used to make membranes in fuel cells. For example, proton exchange membrane fuel cell (PEMFC) and direct methanol/ethanol fuel cell (DEFC/DMFC).

In addition, GO can be used as an electron donor to reduce precious metal ions without the addition of reductants and surfactants. This causes the growth of metal nanoparticles on the GO surface with high dispersion, uniformity and purity [83]. Noble metals with small sizes that have a large surface area and many angles can show good electrocatalytic abilities in fuel cells. For example, mono-dispersed GO mixed with gold nanoparticles was synthesized by a redox reaction between $AuCl_4$ and GO which resulted in high electrocatalytic activity and electron pathway leading to oxygen reduction reactions. A more interesting result was reported that the 3D GO/carbon sphere supported by silver nanocomposite produced using GO as a reductant showed a significant increase in activity for the oxygen reduction reaction in alkaline media [84]. The 3D structure is useful in driving transport in the catalytic layer and facilitating reactant access to the active site.

GO containing conjugate $\pi$ bonds were also attractive as a substitute for expensive noble metal materials. The embedding of GO into the polymer matrix can increase the conductivity and reduce membrane fouling. Polymer GO composites can be synthesized by hydrogen bonding and the epoxy ring opening reaction uses amines to form new C-N bonds. Which can function as an active for the electrocatalytic reduction of $O_2$ to $H_2O$ [85].

## 4. Conclusions

Graphene oxide is a monolayer of carbon atom decorated by oxygen functional group. Abundant and non-used biomass wastes are promising candidate for carbon source to fabricate graphene oxide due to their highly carbon content. Carbonization process is the key step to produce biomass carbon as precursor. Modified Hummer's method is a facile and simple chemical wet process that can be served to synthesis graphene oxide form biomass waste. XRD pattern showed the Graphene Oxide had successfully produced from biomass waste. Graphene oxide plays important role in the advance science and technology today. GO can be applied as starting material to fabricate electrical and optical device such as cathode of battery and supercapacitor, transparent electrode, field effect device, electrical sensor, flexible electronics, photovoltaic device, electrochromic device, photocatalytic, and fuel cell. Numerous researches continuously report advance application of graphene oxide that depict advance graphene oxide-based technology in the future.

## Acknowledgements

The authors would like to thank to Directorate of Research, Technology, and Community Service, Ministry of Education, Culture, Research and Technology of the Republic of Indonesia, through grant: *Penelitian Dasar Kompetitif Nasional*, Contract No. 197/E5/PG.02.00.PT/2022.

## Conflict of interest

The authors declare no conflict of interest.

## Author details

Ramli Ramli* and Rahmat Hidayat
Nanoscience and Nanotechnology Research Group, Department of Physics, Universitas Negeri Padang, Padang, Indonesia

*Address all correspondence to: ramli@fmipa.unp.ac.id

## References

[1] Zhao J, Liu L, Li F. Graphene Oxide: Physics and Applications. New York: Springer; 2015. p. 3. DOI: 10.1007/978-3-662-44829-8

[2] Brodie BC. On the atomic weight of graphite. Philosophical Transactions of the Royal Society of London. Series B, Biological Sciences. 1859;**149**:249-259. DOI: 10.1098/rstl.1859.0013

[3] Dreyer DR, Park S, Bielawski CW, Ruoff RS. The chemistry of graphene oxide. Chemical Society Reviews. 2010;**39**:228-240. DOI: 10.1039/B917103G

[4] Ray SC. Applications of Graphene and Graphene-Oxide Based Nanomaterials. Waltham: William Andrew; 2015. p. 39. DOI: 10.1016/C2014-0-02615-9

[5] Stankovich S, Piner RD, Nguyen ST, Ruoff RS. Synthesis and exfoliation of isocyanate-treated graphene oxide nanoplatelets. Carbon. 2006;**44**:3342-3347. DOI: 10.1016/j.carbon.2006.06.004

[6] Zhu Y, Stoller MD, Cai W, Velamakanni A, Piner RD, Chen D, et al. Exfoliation of graphite oxide in propylene carbonate and thermal reduction of the resulting graphene oxide platelets. ACS Nano. 2010;**4**:1227-1233. DOI: 10.1021/nn901689k

[7] Staudenmaier L. *Verfahren zur Darstellung der Graphitsäure* (Method for the preparation of graphitic acid). Berichte der Deutschen Chemischen Gesellschaft. 1898;**31**:1481-1487. DOI: 10.1002/cber.18980310237

[8] Hummers WS, Offeman RE. Preparation of graphitic oxide. Journal of the American Chemical Society. 1958;**80**:1339. DOI: 10.1021/ja01539a017

[9] Ranjan P, Agrawal S, Sinha A, Rao TR, Balakrishnan J, Thakur AD. A low cost non-explosive synthesis of graphene oxide for scalable applications. Scientific Reports. 2018;**8**:12007. DOI: 10.1038/s41598-018-30613-4

[10] Marcano DC, Kosynkin DV, Berlin JM, Sinitskii A, Sun Z, Slesarev A, et al. Improved synthesis of graphene oxide. ACS Nano. 2010;**4**:4806-4814. DOI: 10.1021/nn1006368

[11] Sierra U, Álvarez P, Blanco C, Granda M, Santamaría R, Menéndez R. Cokes of different origin as precursors of graphene oxide. Fuel. 2016;**166**:400-403. DOI: 10.1016/j.fuel.2015.10.112

[12] Smith AT, LaChance AM, Zeng S, Liu B, Sun L. Synthesis, properties, and applications of graphene oxide/reduced graphene oxide and their nanocomposites. Nano Materials Science. 2019;**1**:31-47. DOI: 10.1016/j.nanoms.2019.02.004

[13] Li F, Jiang X, Zhao J, Zhang S. Graphene oxide: A promising nanomaterial for energy and environmental applications. Nano Energy. 2015;**16**:488-515. DOI: 10.1016/j.nanoen.2015.07.014

[14] Hu BB, Wang K, Wu L, Yu SH, Antonietti M, Titirici MM. Engineering carbon materials from the hydrothermal carbonization process of biomass. Advanced Materials. 2010;**22**:813-828. DOI: 10.1002/adma.200902812

[15] Miao M, Zuo S, Zhao Y, Wang Y, Xia H, Tan C, et al. Selective oxidation rapidly decomposes biomass-based activated carbons into graphite-like crystallites. Carbon. 2018;**40**:504-507. DOI: 10.1016/j.carbon.2018.09.018

[16] Bishnu PT, Huimin L, Phillip H, Harry MM, John RD, Sheng D. Low-cost transformation of biomass-derived carbon to high-performing nano-graphite via low-temperature electrochemical graphitization. ACS Applied Materials & Interfaces. 2021;**13**:4393-4401. DOI: 10.1021/acsami.0c19395

[17] Wang K, Wang Y, Wang C, Xia Y. Graphene oxide assisted solvothermal synthesis of LiMnPO4 naonplates cathode materials for lithium ion batteries. Electrochimica Acta. 2014;**146**:8-14. DOI: 10.1016/J.ELECTACTA.2014.09.032

[18] Zhao B, Wang Z, Chen L, Yang Y, Chen F, Gao Y, et al. $LiMnPO_4$/graphene nanocomposites with high electrochemical performance for lithium-ion batteries. Huagong Xuebao/CIESC Journal. 2016;**67**(11):4779-4786. DOI: 10.11949/j.issn.0438-1157.20160651

[19] Jiang Y, Liu R, Xu W, Jiao Z, Wu M, Chu Y, et al. A novel graphene modified $LiMnPO_4$ as a performance-improved cathode material for lithium-ion batteries. Journal of Materials Research. 2013;**28**(18):2584-2589. DOI: 10.1557/jmr.2013.235

[20] Reddy Channu VS, Ravichandran D, Rambabu B, Holze R. Carbon and functionalized graphene oxide coated vanadium oxide electrodes for lithium ion batteries. Applied Surface Science. Jun. 2014;**305**:596-602. DOI: 10.1016/J.APSUSC.2014.03.140

[21] Yu F, Zhang L, Lai L, Zhu M, Guo Y, Qi P, et al. High electrochemical performance of $LiFePO_4$ cathode material via in-situ microwave exfoliated graphene oxide. Electrochimica Acta. 2015;**151**:240-248. DOI: 10.1016/J.ELECTACTA.2014.11.014

[22] Zhu YG, Wang Y, Xie J, Cao G-S, Zhu T-J, Zhao X, et al. Effects of graphene oxide function groups on $SnO_2$/graphene nanocomposites for lithium storage application. Electrochimica Acta. 2015;**154**:338-344. DOI: 10.1016/J.ELECTACTA.2014.12.065

[23] Song H, Li N, Cui H, Wang C. Enhanced capability and cyclability of $SnO_2$–graphene oxide hybrid anode by firmly anchored $SnO_2$ quantum dots. Journal of Materials Chemistry A. 2013;**1**(26):7558-7562. DOI: 10.1039/c3ta11442b

[24] Bhaskar A, Deepa M, Ramakrishna M, Rao TN. Poly(3,4-ethylenedioxythiophene) sheath over a $SnO_2$ hollow spheres/graphene oxide hybrid for a durable anode in Li-Ion batteries. Journal of Physical Chemistry C. 2014;**118**(14):7296-7306. DOI: 10.1021/jp412038y

[25] Reddy MJK, Ryu SH, Shanmugharaj AM. Synthesis of $SnO_2$ pillared carbon using long chain alkylamine grafted graphene oxide: An efficient anode material for lithium ion batteries. Nanoscale. 2016;**8**(1):471-482. DOI: 10.1039/c5nr06680h

[26] Wu H, Huang Y, Zong M, Ding X, Ding J, Sun X. Electrostatic self-assembly of graphene oxide wrapped sulfur particles for lithium-sulfur batteries. Materials Research Bulletin. 2015;**64**:12-16. DOI: 10.1016/j.materresbull.2014.12.036

[27] Rong J, Ge M, Fang X, Zhou C. Solution ionic strength engineering as a generic strategy to coat graphene oxide (GO) on various functional particles and its application in high-performance lithium-sulfur (Li-S) batteries. Nano Letters. 2014;**14**(2):473-479. DOI: 10.1021/nl403404v

[28] Xiao M, Huang M, Zeng S, Han D, Wang S, Sun L, et al. Sulfur@ graphene oxide core-shell particles as a rechargeable lithium-sulfur battery cathode material with high cycling stability and capacity. RSC Advances. 2013;**3**(15):4914-4916. DOI: 10.1039/ c3ra00017f

[29] Ji L, Rao M, Zheng H, Zhang L, Li Y, Duan W, et al. Graphene oxide as a sulfur immobilizer in high performance lithium/sulfur cells. Journal of the American Chemical Society. 2011;**133**(46):18522-18525. DOI: 10.1021/ ja206955k

[30] Liu S, Xie K, Li Y, Chen Z, Hong X, Zhou L, et al. Graphene oxide wrapped hierarchical porous carbon-sulfur composite cathode with enhanced cycling and rate performance for lithium sulfur batteries. RSC Advances. 2015;**5**(8):5516-5522. DOI: 10.1039/ c4ra12393j

[31] Zhou W, Chen H, Yu Y, Wang D, Cui Z, DiSalvo FJ, et al. Amylopectin wrapped graphene oxide/sulfur for improved cyclability of lithium-sulfur battery. ACS Nano. 2013;**7**(10):8801-8808. DOI: 10.1021/nn403237b

[32] Zhang J, Yang N, Yang X, Li S, Yao J, Cai Y. Hollow sulfur@graphene oxide core-shell composite for high-performance Li-S batteries. Journal of Alloys and Compounds. 2015;**650**:604-609. DOI: 10.1016/j.jallcom.2015.08.050

[33] Sun C, Shi L, Fan C, Fu X, Ren Z, Qian G, et al. Bowl-like sulfur particles wrapped by graphene oxide as cathode material of lithium-sulfur batteries. RSC Advances. 2015;**5**(36):28832-28835. DOI: 10.1039/c5ra00744e

[34] Huang X, Shi K, Yang J, Mao G, Chen J. MnO2-GO double-shelled sulfur (S@MnO2@GO) as a cathode for Li-S batteries with improved rate capability and cyclic performance. Journal of Power Sources. 2017;**356**:72-79. DOI: 10.1016/j. jpowsour.2017.04.065

[35] Hwa Y, Seo HK, Yuk J-M, Cairns EJ. Freeze-dried sulfur-graphene oxide-carbon nanotube nanocomposite for high sulfur-loading lithium/sulfur cells. Nano Letters. 2017;**17**(11):7086-7094. DOI: 10.1021/acs.nanolett.7b03831

[36] Du X, Zhang X, Guo J, Zhao S, Zhang F. Hierarchical sulfur confinement by graphene oxide wrapped, walnut-like carbon spheres for cathode of Li-S battery. Journal of Alloys and Compounds. 2017;**714**:311-317. DOI: 10.1016/j.jallcom.2017.04.258

[37] Tian Y, Sun Z, Zhang Y, Wang X, Bakenov Z, Yin F. Micro-spherical sulfur/ graphene oxide composite via spray drying for high performance lithium sulfur batteries. Nanomaterials. 2018;**8**(50):1-12. DOI: 10.3390/ nano8010050

[38] Tian Y, Yu Z, Cao L, Zhang XL, Sun C, Wang D-W. Graphene oxide: An emerging electromaterial for energy storage and conversion. Journal of Energy Chemistry. 2021;**55**:323-344. DOI: 10.1016/j.jechem.2020.07.006

[39] Zhang LL, Zhou R, Zhao XS. Graphene-based materials as supercapacitor electrodes. Journal of Materials Chemistry. 2010;**20**(29):5983-5992. DOI: 10.1039/c000417k

[40] Yang P, Mai W. Flexible solid-state electrochemical supercapacitors. Nano Energy. 2014;**8**:274-290. DOI: 10.1016/J. NANOEN.2014.05.022

[41] Karthika P, Rajalakshmi N, Dhathathreyan KS. Functionalized

Exfoliated Graphene Oxide as Supercapacitor Electrodes. Soft Nanoscience Letters. 2012;**2**(4):59-66. DOI: 10.4236/snl.2012.24011

[42] Dighe AB, Dubal DP, Holze R. Screen printed asymmetric supercapacitors based on LiCoO2 and graphene oxide. Zeitschrift für Anorganische und Allgemeine Chemie. 2014;**640**(14):2852-2857. DOI: 10.1002/zaac.201400319

[43] Dai K, Lu L, Liang C, Dai J, Liu Q, Zhang Y, et al. In situ assembly of $MnO_2$ nanowires/graphene oxide nanosheets composite with high specific capacitance. Electrochimica Acta. 2014;**116**:111-117. DOI: 10.1016/j.electacta.2013.11.036

[44] Chen S, Zhu J, Wu X, Han Q, Wang X. Graphene oxide–MnO2 nanocomposites for supercapacitors. ACS Nano. 2010;**4**(5):2822-2830. DOI: 10.1021/nn901311t

[45] Jafta CJ, Nkosi F, le Roux L, Mathe MK, Kebede M, Makgopa K, et al. Manganese oxide/graphene oxide composites for high-energy aqueous asymmetric electrochemical capacitors. Electrochimica Acta. 2013;**110**:228-233. DOI: 10.1016/j.electacta.2013.06.096

[46] Liu Y, Yan D, Li Y, Wu Z, Zhuo R, Li S, et al. Manganese dioxide nanosheet arrays grown on graphene oxide as an advanced electrode material for supercapacitors. Electrochimica Acta. 2014;**117**:528-533. DOI: 10.1016/j.electacta.2013.11.121

[47] Gund GS, Dubal DP, Patil BH, Shinde SS, Lokhande CD. Enhanced activity of chemically synthesized hybrid graphene oxide/Mn3O4 composite for high performance supercapacitors. Electrochimica Acta. 2013;**92**:205-215. DOI: 10.1016/j.electacta.2012.12.120

[48] Wu M-S, Lin Y-P, Lin C-H, Lee J-T. Formation of nano-scaled crevices and spacers in NiO-attached graphene oxide nanosheets for supercapacitors. Journal of Materials Chemistry. 2012;**22**(6):2442-2448. DOI: 10.1039/c1jm13818a

[49] Wang A, Wang H, Zhang S, Mao C, Song J, Niu H, et al. Controlled synthesis of nickel sulfide/graphene oxide nanocomposite for high-performance supercapacitor. Applied Surface Science. 2013;**282**:704-708. DOI: 10.1016/j.apsusc.2013.06.038

[50] Nicasio-Collazo J, Maldonado J-L, Salinas-Cruz J, Barreiro-Argüelles D, Caballero-Quintana I, Vazquez-Espinosa C, et al. Functionalized and reduced graphene oxide as hole transport layer and for use in ternary organic solar cell. Optical Materials. 2019;**98**:109434. DOI: 10.1016/j.optmat.2019.109434

[51] Simon P, Gogotsi Y. Materials for electrochemical capacitors. Nature Materials. 2008;**7**(11):845-854. DOI: 10.1038/nmat2297

[52] Zhang LL, Zhao S, Tian XN, Zhao XS. Layered graphene oxide nanostructures with sandwiched conducting polymers as supercapacitor electrodes. Langmuir. 2010;**26**(22):17624-17628. DOI: 10.1021/la103413s

[53] Liu Y, Deng R, Wang Z, Liu H. Carboxyl-functionalized graphene oxide–polyaniline composite as a promising supercapacitor material. Journal of Materials Chemistry. 2012;**22**(27):13619-13624. DOI: 10.1039/C2JM32479B

[54] Xu G, Wang N, Wei J, Lv L, Zhang J, Chen Z, et al. Preparation of graphene oxide/polyaniline nanocomposite with assistance of supercritical carbon dioxide for supercapacitor electrodes. Industrial and Engineering Chemistry Research.

2012;**51**(44):14390-14398. DOI: 10.1021/ie301734f

[55] Wang H, Hao Q, Yang X, Lu L, Wang X. Graphene oxide doped polyaniline for supercapacitors. Electrochemistry Communications. 2009;**11**(6):1158-1161. DOI: 10.1016/j.elecom.2009.03.036

[56] Luo Z, Zhu L, Zhang H, Tang H. Polyaniline uniformly coated on graphene oxide sheets as supercapacitor material with improved capacitive properties. Materials Chemistry and Physics. 2013;**139**(2):572-579. DOI: 10.1016/j.matchemphys.2013.01.059

[57] Wang H, Hao Q, Yang X, Lu L, Wang X. Effect of graphene oxide on the properties of its composite with polyaniline. ACS Applied Materials & Interfaces. 2010;**2**(3):821-828. DOI: 10.1021/am900815k

[58] Pendashteh A, Mousavi MF, Rahmanifar MS. Fabrication of anchored copper oxide nanoparticles on graphene oxide nanosheets via an electrostatic coprecipitation and its application as supercapacitor. Electrochimica Acta. 2013;**88**:347-357. DOI: 10.1016/j.electacta.2012.10.088

[59] Xu J, Wang K, Zu S-Z, Han B-H, Wei Z. Hierarchical nanocomposites of polyaniline nanowire arrays on graphene oxide sheets with synergistic effect for energy storage. ACS Nano. 2010;**4**(9):5019-5026. DOI: 10.1021/nn1006539

[60] Xu Y, Liu J. Graphene as transparent electrodes: fabrication and new emerging applications. Small. 2016;**12**(11):1400-1419

[61] Becerril HA, Mao J, Liu Z, Stoltenberg RM, Bao Z, Chen Y. Evaluation of solution-processed reduced graphene oxide films as transparent conductors. ACS Nano. 2008;**2**(3):463-470. DOI: 10.1021/nn700375n

[62] Aimon AH, Hidayat R, Rahmawati D, Sutarto R, Permatasari FA, Iskandar F. Facile deposition of reduced graphene oxide-based transparent conductive film with microwave assisted method. Thin Solid Films. 2019;**692**:137618. DOI: 10.1016/j.tsf.2019.137618

[63] Trung TQ, Duy LT, Ramasundaram S, Lee N-E. Transparent, stretchable, and rapid-response humidity sensor for body-attachable wearable electronics. Nano Research. 2017;**10**(6):2021-2033. DOI: 10.1007/s12274-016-1389-y

[64] Zhu Y, Murali S, Cai W, Li X, Suk JW, Potts JR, et al. Graphene and graphene oxide: Synthesis, properties, and applications. Advanced Materials. 2010;**22**:3906-3924

[65] Novoselov KS, Geim AK, Morozov SV, Jiang D, Zhang Y, Dubonos SV, et al. Electric field effect in atomically thin carbon films. Science (80). 2004;**306**(5696):666-669. DOI: 10.1126/science.1102896

[66] Yamaguchi H, Murakami K, Eda G, Fujita T, Guan P, Wang W, et al. Field emission from atomically thin edges of reduced graphene oxide. ACS Nano. 2011;**5**(6):4945-4952. DOI: 10.1021/nn201043a

[67] Joung D, Chunder A, Zhai L, Khondaker SI. High yield fabrication of chemically reduced graphene oxide field effect transistors by dielectrophoresis. Nanotechnology. 2010;**21**(16):165202

[68] Kim T-W, Gao Y, Acton O, Yip H-L, Ma H, Chen H, et al. Graphene oxide nanosheets based organic

field effect transistor for nonvolatile memory applications. Applied Physics Letters. 2010;**97**:0233101-0233103. DOI:10.1063/1.3464292

[69] Jung I, Dikin DA, Piner RD, Ruoff RS. Tunable electrical conductivity of individual graphene oxide sheets reduced at 'low' temperatures. Nano Letters. 2008;**8**(12):4283-4287

[70] Robinson JT, Perkins FK, Snow ES, Wei Z, Sheehan PE. Reduced graphene oxide molecular sensors. Nano Letters. 2008;**8**(10):3137-3140. DOI: 10.1021/nl8013007

[71] Robinson JA, Snow ES, Bădescu ŞC, Reinecke TL, Perkins FK. Role of defects in single-walled carbon nanotube chemical sensors. Nano Letters. 2006;**6**(8):1747-1751

[72] Lin Y-M, Avouris P. Strong suppression of electrical noise in bilayer graphene nanodevices. Nano Letters. 2008;**8**(8):2119-2125. DOI: 10.1021/nl080241l

[73] Eda G, Fanchini G, Chhowalla M. Large-area ultrathin films of reduced graphene oxide as a transparent and flexible electronic material. Nature Nanotechnology. 2008;**3**(5):270-274

[74] Lee SH, Kim HW, Hwang JO, Lee WJ, Kwon J, Bielawski C, et al. Three-dimensional self-assembly of graphene oxide platelets into mechanically flexible macroporous carbon films. Angewandte Chemie. 2010;**122**(52):10282-10286. DOI: 10.1002/ange.201006240

[75] Su Q, Pang S, Alijani V, Li C, Feng X, Müllen K. Composites of graphene with large aromatic molecules. Advanced Materials. 2009;**21**(31):3191-3195

[76] Wu J, Agrawal M, Becerril HA, Bao Z, Liu Z, Chen Y, et al. Organic light-emitting diodes on solution-processed graphene transparent electrodes. ACS Nano. 2010;**4**(1):43-48. DOI: 10.1021/nn900728d

[77] Palenzuela J, Vinuales A, Odriozola I, Cabanero G, Grande HJ, Ruiz V. Flexible viologen electrochromic devices with low operational voltages using reduced graphene oxide electrodes. ACS Applied Materials & Interfaces. 2014;**6**(16):14562-14567

[78] Wang X, Zhi L, Müllen K. Transparent, conductive graphene electrodes for dye-sensitized solar cells. Nano Letters. 2008;**8**(1):323-327

[79] Wu J, Becerril HA, Bao Z, Liu Z, Chen Y, Peumans P. Organic solar cells with solution-processed graphene transparent electrodes. Applied Physics Letters. 2008;**92**(26):237

[80] Sheng K, Bai H, Sun Y, Li C, Shi G. Layer-by-layer assembly of graphene/polyaniline multilayer films and their application for electrochromic devices. Polymer (Guildf). 2011;**52**(24):5567-5572. DOI: 10.1016/j.polymer.2011.10.001

[81] Yeh T-F, Chen S-J, Yeh C-S, Teng H. Tuning the electronic structure of graphite oxide through ammonia treatment for photocatalytic generation of H2 and O2 from water splitting. Journal of Physical Chemistry C. 2013;**117**(13):6516-6524

[82] Zhang J, Yu J, Jaroniec M, Gong JR. Noble metal-free reduced graphene oxide-$Zn_xCd_{1-x}S$ nanocomposite with enhanced solar photocatalytic H2-production performance. Nano Letters. 2012;**12**(9):4584-4589. DOI: 10.1021/nl301831h

[83] Chen X, Wu G, Chen J, Chen X, Xie Z, Wang X. Synthesis of 'clean' and well-dispersive Pd nanoparticles with

excellent electrocatalytic property on graphene oxide. Journal of the American Chemical Society. 2011;**133**(11):3693-3695. DOI: 10.1021/ja110313d

[84] Yuan L, Jiang L, Liu J, Xia Z, Wang S, Sun G. Facile synthesis of silver nanoparticles supported on three dimensional graphene oxide/carbon black composite and its application for oxygen reduction reaction. Electrochimica Acta. 2014;**135**:168-174. DOI: 10.1016/j.electacta.2014.04.137

[85] Naveen MH, Noh H-B, Al Hossain MS, Kim JH, Shim Y-B. Facile potentiostatic preparation of functionalized polyterthiophene-anchored graphene oxide as a metal-free electrocatalyst for the oxygen reduction reaction. Journal of Materials Chemistry A. 2015;**3**(10):5426-5433. DOI: 10.1039/C4TA06774F